ENERGY POLICIES, POLITICS AND PRICES

VULNERABILITIES OF THE U.S. TO AN ELECTROMAGNETIC PULSE ATTACK

THREAT ASSESSMENTS AND MITIGATION RECOMMENDATIONS

ENERGY POLICIES, POLITICS AND PRICES

Additional books in this series can be found on Nova's website under the Series tab.

Additional e-books in this series can be found on Nova's website under the e-book tab.

ENERGY POLICIES, POLITICS AND PRICES

VULNERABILITIES OF THE U.S. TO AN ELECTROMAGNETIC PULSE ATTACK

THREAT ASSESSMENTS AND MITIGATION RECOMMENDATIONS

MARYANNE SCHNEIDER
EDITOR

nova publishers
New York

Library of Congress Cataloging-in-Publication Data

ISBN: 978-1-63484-477-2

Published by Nova Science Publishers, Inc. † New York

CONTENTS

PREFACE

The physical and social fabric of the United States is sustained by a system of systems; a complex and dynamic network of interlocking and interdependent infrastructures ("critical national infrastructures") whose harmonious functioning enables the myriad actions, transactions, and information flow that undergird the orderly conduct of civil society in this country. The vulnerability of these infrastructures to threats — deliberate, accidental, and acts of nature — is the focus of greatly heightened concern in the current era, a process accelerated by the events of 9/11 and recent hurricanes, including Katrina and Rita. This book presents the results of the Commission's assessment of the effects of a high altitude electromagnetic pulse (EMP) attack on our critical national infrastructures and provides recommendations for their mitigation. Moreover, as of July 2015, the Department of Homeland Security (DHS) reported taking several actions that could help address electromagnetic threats to the electric grid. This book also addresses the extent to which DHS has: taken action to address recommendations from the 2008 EMP Commission Report and coordinated with other principal federal agencies, such as DOE and industry stakeholders to mitigate risks to the electric grid from electromagnetic threats.

In: Vulnerabilities of the U.S. to an Electromagnetic Pulse Attack ISBN: 978-1-63484-477-2
Editor: Maryanne Schneider

Chapter 1

REPORT OF THE COMMISSION TO ASSESS THE THREAT TO THE UNITED STATES FROM ELECTROMAGNETIC PULSE (EMP) ATTACK*

Critical National Infrastructures

SECTION 1. INFRASTRUCTURE COMMONALITIES

The physical and social fabric of the United States is sustained by a system of systems; a complex and dynamic network of interlocking and interdependent infrastructures ("critical national infrastructures") whose harmonious functioning enables the myriad actions, transactions, and information flow that undergird the orderly conduct of civil society in this country. The vulnerability of these infrastructures to threats — deliberate, accidental, and acts of nature — is the focus of heightened concern in the current era, a process accelerated by the events of 9/11 and recent hurricanes, including Katrina and Rita.

This volume focuses on a description of the potential vulnerabilities of our critical national infrastructures to electromagnetic pulse (EMP) insult, and to that end, the chapters in this document deal individually with the EMP threat to each critical infrastructure separately. However, to set the stage for understanding the potential threat under conditions in which all infrastructures are under simultaneous attack, it is important to realize that the vulnerability of the whole — of all the highly interlocked critical infrastructures — may be greater than the sum of the vulnerability of its parts. The whole is a highly complex system of systems whose exceedingly dynamic and coordinated activity is enabled by the growth of technology and where failure within one individual infrastructure may not remain isolated but, instead, induce cascading failures into other infrastructures.

It is also important to understand that not only mutual interdependence, and hence new vulnerabilities, may be enabled by technology advances, but also technologies that have facilitated this growing interdependence may be common across the many individual

* This is an edited, reformatted and augmented version of a report issued April 2008.

infrastructures. In particular, the Commission thought it important to single out the growth and common infrastructural infiltration of one particular transformative technology, the development of automated monitoring and control systems — the ubiquitous robots of the modern age known as Supervisory Control and Data Acquisition (SCADA) systems.

This opening Section thus focuses on a more detailed description of these two aspects of modern infrastructures, control systems and mutual interdependence, that are common to all and which the Commission believes provide context and insight for understanding sources of vulnerability in all the Nation's infrastructures to EMP attack.

SCADA Systems

Introduction

SCADAs have emerged as critical and growing elements of a quietly unfolding industrial revolution spurred by the computer age. The accelerating penetration of SCADA systems, along with their electronic cousins, digital control systems (DCS) and programmable logic controllers (PLC), as critical elements in every aspect of every critical infrastructure in the Nation, is both inevitable and inexorable. While conferring economic benefit and enormous new operational agility, the growing dependence of our infrastructures on these omnipresent control systems represents a new vector of vulnerability in the evolving digital age of the 21st century, such as cyber security. Such issues remain as a matter for high-level concern and attention today. High-altitude EMP focuses our attention toward another potential vulnerability of these systems, and one with potentially vastly expanded consequences.

What Is a SCADA?

SCADAs are electronic control systems that may be used for data acquisition and control over large and geographically distributed infrastructure systems. They find extensive use in critical infrastructure applications such as electrical transmission and distribution, water management, and oil and gas pipelines. SCADA technology has benefited from several decades of development. It has its genesis in the telemetry systems used by the railroad and aviation industries.

In November 1999, San Diego County Water Authority and San Diego Gas and Electric companiesexperienced severe electromagnetic interference to their SCADA wireless networks. Both companies found themselves unable to actuate critical valve openings and closings under remote control of the SCADA electronic systems. This inability necessitated sending technicians to remote locations to manually open and close water and gas valves, averting, in the words of a subsequent letter of complaint by the San Diego County Water Authority to the Federal Communications Commission, a potential "catastrophic failure" of the aqueduct system. The potential consequences of a failure of this 825 million gallon per day flow rate system ranged from "spilling vents at thousands of gallons per minute to aqueduct rupture with ensuing disruption of service, severe flooding, and related damage to private and public property." The source of the SCADA failure was later determined to be radar operated on a ship 25 miles off the coast of San Diego.

The physical form of a SCADA may differ from application to application and from one industry to another, but generally they all share certain generic commonalities. A SCADA system physically bears close resemblance to the internals of a generic desktop personal computer. Typically, it might contain familiar-appearing circuit boards, chips of various sorts, and cable connectors to the external world. The cable connectors, in turn, may be connected, perhaps quite remotely, to various sensor systems that are the SCADA's eyes and ears, as well as electronic control devices by which the SCADA may issue commands that adjust system performance. **Figure 1-1** provides an example of a particular SCADA controller that is representative of many such systems.

One major function of a SCADA — the data acquisition part of the acronym — is to provide a capability to automatically and remotely monitor the operating state of a physical system. It accomplishes this monitoring by providing an ongoing reporting of parameters that either characterize the system's performance, such as voltage or currents developed in an electric power plant, flow volume in a gas pipeline, and net electrical power delivered or received by a regional electrical system, or by monitoring environmental parameters such as temperature in a nuclear power plant and sending an alarm when prescribed operating conditions are exceeded.

The supervisory control function of a SCADA reflects the ability of these devices to actively control the operation of the system by adjusting its output. For example, should an electrical generating plant fail through loss of a critical hardware component or industrial accident, the monitoring SCADA will detect the loss, issue an alert to the appropriate authorities, and issue commands to other generating plants under its control to increase their power output to match the load again. All these actions take place automatically, within seconds, and without a human being involved in the immediate control loop.

Figure 1-1. Typical SCADA Architecture.

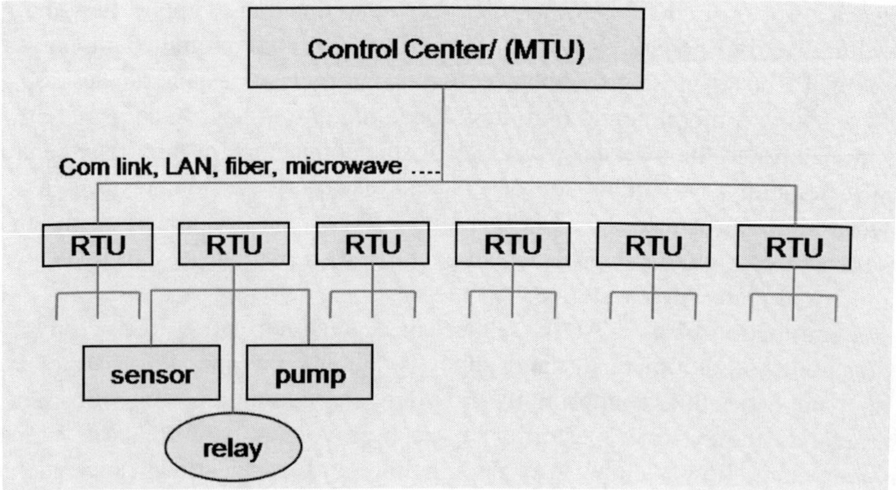

Figure 1-2. Generic SCADA Architecture.

A typical SCADA architecture for the electric power industry may consist of a centralized computer — the master terminal unit (MTU) — communicating through many remote terminal unit (RTU) subsystems, as illustrated in **figure 1-2**. The RTUs are used in remote, unmanned locations where data acquisition and control tasks must be performed. Examples of typical RTU data acquisition actions include processing signals from sensors such as thermocouples, voltage sensors, or power meters and reporting the state of equipment such as switch and circuit breaker positions. Typical control actions include starting and stopping motors and controlling valves and circuit breakers.

DCSs share many functional and physical hardware similarities with SCADA systems. A DCS typically will be used to control automated processes at a single location, such as an oil refinery or a chemical plant. In contrast, a SCADA typically might be sited in an environment with dispersed assets where real-time situational awareness from remote locations is a key element of centralized control. Most DCS installations control complex, dynamic systems that would be difficult or impossible to control in a safe or economical manner using only manual control.

Even a relatively straightforward process such as electrical power generation using a conventional steam cycle requires highly complex systems to maximize efficiency, while maintaining safety and environmental protection. For example, control systems in a steam generating plant would include parameters such as generator speed, generator lubrication oil pressure, excitation current and voltage output, feed water pressure and boiler steam drum level, and air box pressure and rate of combustion.

Upset of these control points has the potential to cause severe physical damage. A case in point is the boiler endpoints of combustion and circulation. Normally, the control system would first reach the endpoint of combustion (limit of air and fuel adding energy into the boiler) and, thus, prevent any thermal damage to the boiler. If the control system is upset, it potentially could reach the endpoint of circulation (maximum rate of steam generation) or endpoint of carryover (maximum rate at which water is *not* carried out of the boiler) before the endpoint of combustion. This situation would cause thermal damage to the boiler tubes or physical damage to steam turbine blades.

Figure 1-3. PLC Switch Actuator.

Normally, a PLC is used to control actuators or monitor sensors and is another piece of hardware that shares many physical similarities to SCADAs and is often found as part of a larger SCADA or DCS system. The SCADA, DCS, and PLC systems all share electronic commonalities and, thus they share intrinsic electronic vulnerabilities as well. SCADA systems, however, tend to be more geographically disposed and exposed; our subsequent discussion focuses on SCADAs. When exposure or unprotected cable connectivity is an issue, the discussion should be considered to pertain to both PLC and DCS as well. See **figure 1-3**.

EMP Interaction with SCADA

SCADA system components by their nature are frequently situated in remote environments and operate without proximate human intervention. Although their critical electronic elements usually are contained within some sort of metallic box, the enclosures' service as a protective Faraday cage is typically minimal. Generally such metallic containers are designed only to provide protection from the elements and a modicum of physical security. They typically are not designed to protect the electronics from high-energy electromagnetic pulses that may infiltrate either from the free field or from the many antennae (cable connections) that may compromise electromagnetic integrity. The major concern for SCADA vulnerability to EMP is focused on the early time E1 component of the EMP signal. This is because, even in the power industry, SCADA systems generally are not directly coupled electrically to the very long cable runs that might be expected to couple to a late-time E3 signal.

To come to grips with the potential vulnerability of our critical national infrastructures caused by a threat to these ubiquitous SCADA control systems, we must first develop a sense of the vulnerability of the underlying hardware components themselves. To this end, the EMP Commission sponsored and funded a series of tests of common SCADA components in a government-owned EMP simulator (see **figure 1-4**). The simulation testing provided an opportunity to observe the interaction of the electromagnetic energy with equipment in an operational mode. Because the simulator did not completely replicate all characteristics of a threat-level EMP environment, observed test results can be related to the system's response in more realistic scenarios through analysis and judgment based on coupling differences between the simulated and real-world cases.

The Commission consulted with experts from industry groups associated with the North American Electric Reliability Corporation (NERC) and by site and market surveys to identify representative control systems for testing. A test matrix was developed that reflected electronic control technologies employed in power generation, power distribution, pipeline distribution, and manufacturing plants. Some test items assessed in this effort are shown in **figure 1-5**.

EMP Simulation Testing

In this section, we provide a brief summary of the results of illuminating electronic control systems in the simulator. The detailed results of the simulation test program are documented separately in reports sponsored by the Commission. In Section 2, we provide a more complete description of the test methodology, in which we discuss testing carried out during assessment of the EMP vulnerability of the electric grid.

Figure 1-4. EMP Simulator with Test Structures and Internal Electronics.

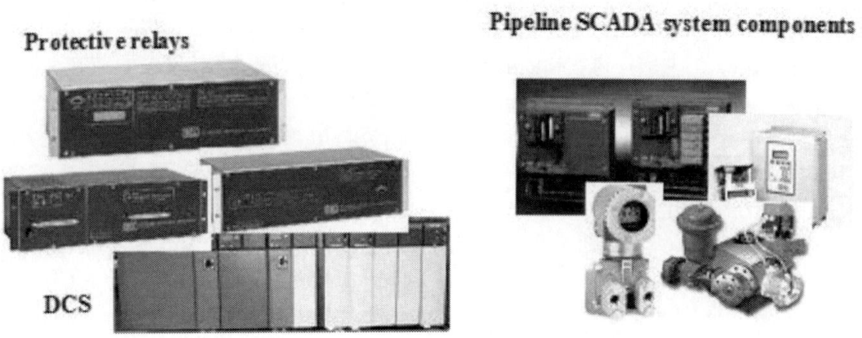

Figure 1-5. Some of the Electronic Control Systems Exposed in Test Facility.

Many of the control systems that we considered achieved operational connectivity through Ethernet cabling. EMP coupling of electrical transients to the cables proved to be an important vulnerability during threat illumination. Because the systems would require manual repair, their full restoration could be a lengthy process. A simple model of four Ethernet cables from a router to four personal computers (PC) was generated to quantify the impact of cable length. The configuration of this model is shown in **figure 1-6**. The results of the analysis indicate that the coupling to the 200 feet of Ethernet line is roughly seven times the transient level on the 25-foot line measured during the test program. The testing and analysis indicate that the electronics could be expected to see roughly 100 to 700 ampere current transients on typical Ethernet cables. Effects noted in the EMP testing occurred at the lower end of this scale.

The bottom line observation at the end of the testing was that every system tested failed when exposed to the simulated EMP environment. The failures were not identical from system to system or within a system. For example, a device with many input-output ports might exhibit degraded performance on one port, physical damage on another, and no effect on a third. Control units might report operating parameters at variance with their post illumination reality or fail to control internal flows. The Commission considered the implications of these multiple simultaneous control system failures to be highly significant as potential contributors to a widespread system collapse.

Impact of SCADA Vulnerabilities on Critical Infrastructures: Historical Insight

Based on the testing and analysis outlined in the previous section, we estimate that a significant fraction of all remote control systems within the EMP-affected area will experience some type of impact. As the test results were briefed to industry experts at NERC and the Argonne National Laboratory, it became apparent that even minor effects noted during the testing could significantly affect the processes and equipment being controlled. Putting together a complete analysis for complex processes associated with infrastructure systems is extremely difficult. Developing the ability to analyze or model these impacts is beyond the scope of this effort.

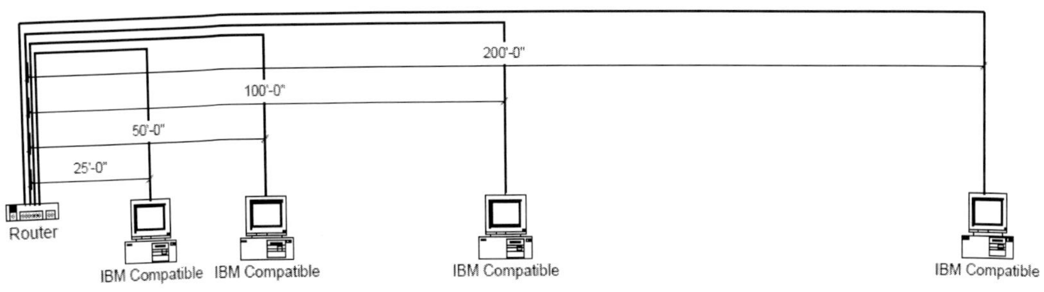

Figure 1-6. Physical Model Used to Quantify Coupling to Different Cable Lengths in a Hypothetical Local Area Network (LAN).

To provide insight into the potential impact of these EMP-induced electronic system malfunctions, one can consider the details of historical events. In these cases, similar (and arguably less severe) system malfunctions have produced consequences in situations that are far too complex to predict beforehand using a model or analysis.

Another important observation is that these incidents are seldom the result of a single factor. Rather they are a combination of unexpected events that, only in hindsight, are easily related to the impact. This is not surprising given the complexity of the systems involved. Before considering the historical database, it is important to remember that historical examples, although important for the insight they provide into the dependence of a functioning modern infrastructure on its automated eyes, ears, and remote controllers, do not adequately capture the scale of the expected EMP scenario. In the latter, it is not one or a few SCADA systems that are malfunctioning (the typical historical scenario), but large numbers — hundreds or even thousands — with some fraction of those rendered permanently inoperable until replaced or physically repaired.

Significant historical events that provide insight into the potential impact of damage or upset to control systems include Hurricane Katrina; the 1996 Western States blackout; the August 14, 2003, Northeast blackout; a geomagnetic storm in 1989; the June 10, 1999, Bellingham pipeline incident; the August 19, 2000, Carlsbad pipeline incident; the July 24, 1994, Pembroke, United Kingdom, refinery incident; and a Netherlands electromagnetic interference (EMI) incident. The following paragraphs discuss the relevance of four of these incidents to an EMP event. The other four incidents — Hurricane Katrina; the Western States blackout; the August 14, 2003, blackout; and the 1989 geomagnetic storm — are described in Section 2, which is dedicated to a discussion of EMP effects on the electric power grid.

Bellingham Pipeline Incident

On June 10, 1999, one of the Olympic pipelines transporting gasoline ruptured in the Whatcom Falls Park area of Bellingham, Washington. About 250,000 gallons of gasoline from the pipeline entered the Hannah and Whatcom Creeks, where the fuel ignited, resulting in three fatalities and eight injuries. In addition, the banks of the creek were destroyed over a 1.5-mile section, and several buildings adjacent to the creek were severely damaged.

Causes included improperly set relief valves, delayed maintenance inspections, and SCADA system discrepancies. The effects all came together at the same time that changes in pipeline operations were occurring. Given the wide area of an EMP, it is conceivable that some of the pipelines affected could also suffer from poor maintenance. The electronic disturbance of an EMP event could be expected to precipitate SCADA failures and the ensuing loss of valve controls.

Carlsbad Pipeline Incident

On August 19, 2000, an explosion occurred on one of three adjacent large natural gas pipelines near Carlsbad, New Mexico, operated by the El Paso Natural Gas Company. The pipelines supply consumers and electric utilities in Arizona and Southern California. Twelve people, including five children, died as a result of the explosion. The explosion left an 86-foot-long crater. After the pipeline failure, the Department of Transportation's Office of Pipeline Safety (OPS) ordered the pipeline to be shut down. The explosion happened because of failures in maintenance and loss of situational awareness, conditions that would be replicated by data acquisition disruptions caused by an EMP event.

Pembroke Refinery Incident

On July 24, 1994, a severe thunderstorm passed over the Pembroke refinery in the United Kingdom. Lightning strikes resulted in a 0.4 second power loss and subsequent power dips throughout the refinery. Consequently, numerous pumps and overhead fin-fan coolers tripped repeatedly, resulting in the main crude column pressure safety valves lifting and major upsets in the process units in other refinery units, including those within the fluid catalytic cracking (FCC) complex.

There was an explosion in the FCC unit and a number of isolated fires continued to burn at locations within the FCC, butamer, and alkylation units. The explosion was caused by flammable hydrocarbon liquid continuously being pumped into a process vessel that, because of a valve malfunction, had its outlet closed. The control valve was actually shut when the control system indicated that it was open. The malfunctioning process control system did not allow the refinery operators to contain the situation.

As a result of this incident, an estimated 10 percent of the total refining capacity in the United Kingdom was lost until this complex was returned to service. The business loss is estimated at $70 million, which reflects 4.5 months of downtime. The disturbances caused by the lightning strikes — power loss and degradation — would also result from an EMP event.

Netherlands EMI Incident

A mishap occurred at a natural gas pipeline SCADA system located about 1 mile from the port of Den Helder, Netherlands, in the late 1980s. A SCADA disturbance caused a catastrophic failure of an approximately 36-inch diameter pipeline, which resulted in a large gas explosion.

This failure was caused by EMI traced to a radar coupling into the wires of the SCADA system. Radio frequency energy caused the SCADA system to open and close at the radar scan frequency, a relay that was, in turn, controlling the position of a large gas flow-control valve. The resulting changes in valve position created pressure waves that traveled down the pipeline and eventually caused the pipeline to fail. This incident shows the potential damage to pipelines from improper control system operations, a condition that could be replicated by an EMP event.

Summary

SCADA systems are vulnerable to EMP insult. The large numbers and widespread reliance on such systems by all of the Nation's critical infrastructures represent a systemic threat to their continued operation following an EMP event. Additionally, the necessity to reboot, repair, or replace large numbers of geographically widely dispersed systems will considerably impede the Nation's recovery from such an assault.

Infrastructures and their Interdependencies

Introduction

All critical national infrastructures are fault tolerant to some degree. Design engineers and system managers are cognizant of, and fully expect, failure of subsystems and individual electrical components. Networks are designed with an expressed goal to avoid single point failures that can bring down the entire system, though in practice the evolved network may be so complicated that no one can guarantee that this design goal has been achieved. Single point failures are anticipated in the design of the systems and engineering solutions of various kinds, including redundancy, rapid repair, replacement, and operational rerouting.

It is important to note, however, that safeguards against single point failures generally depend on the proper functioning of the rest of the national infrastructure, a plausible assumption for high-reliability infrastructure systems when they experience random, uncorrelated single point failures.

Planning for multiple failures, particularly when they are closely correlated in time, is much less common. It is safe to say that no one has planned for, and few have even imagined, a scenario with the loss of hundreds or even thousands of nodes across all the critical national infrastructures, all simultaneously. That, however, is precisely the circumstance contemplated by an EMP attack scenario.

The ability to predict the consequences of failure within a critical infrastructure will require the use of reliable modeling and simulation tools. Some tools exist for the individual infrastructures and serve as either planning tools, real-time control models, or operational support elements to allocate or control resources during network outages and restoration activities. They are generally validated within the parameter space of normal operating experience and concern, and they serve their purposes well. But it is also recognized that the systems being modeled are so complex that currently available modeling tools cannot capture the full richness of potential system responses to all possible network configurations and operating states.

> "We have produced designs so complicated that we cannot possibly anticipate all the possible interactions of the inevitable failures; we add safety devices that are deceived or avoided or defeated by hidden paths in the systems." Charles Perrow, *Normal Accidents*.

Thus, for example, on the order of once a decade or so, portions of the national power grid will experience an unpredicted major disruption with failures cascading through some of the network pathways. Following the major Northeast power blackout of August 14, 2003, analysts continued to debate the cause of the disruption. The sophisticated, relatively mature, and operationally deployed modeling and simulation tools have not been able to replicate unambiguously the observed events of August 14.

The scenarios envisioned by an EMP attack involve potential failures distributed across a wide geographical extent. These include multiple combinations of node failures, a condition that generally is outside the parameter space of validation of extant system models and poses a severe challenge to predicting the subsequent evolution of the infrastructure response. The response of critical national infrastructures to an EMP attack is precisely the subject of many sections of this report. But there is a particular aspect of modeling infrastructures that is even less well developed and whose particular relevance to the EMP scenario stresses the current state of the simulation art to produce a high-fidelity simulation. That aspect is the interaction among the different infrastructures. Particularly difficult to anticipate and to capture in simulations are situations in which the occurrence of simultaneous failures can bring into play dormant and hitherto hidden interaction pathways in which a destructively synergistic amplification of failure, normally locally contained, may be propagated through the network at large.

Charles Perrow[1] in particular has drawn attention to these types of failures, which he has termed *normal accidents* and which are posited as an inherent property of any tightly coupled system once a threshold of complexity has been passed. The Commission believes that, given sufficient priority, time and resources, complex interdependent models can be developed to guide future assessments of the U.S. national infrastructure to EMP attack and to guide investment decisions on how best to protect our infrastructures.

Complex Interactions

Various lists are in circulation that identify the critical infrastructures. The EMP Commission has chosen to address the following areas in separate sections of this Commission report:

- Electric power
- Telecommunications
- Banking and finance
- Petroleum and natural gas
- Transportation
- Food
- Water
- Emergency services
- Space
- Government

The separation of these infrastructures into different domains tends to obscure the real interdependencies that sustain the effectiveness and daily operation of each one.

As a simple example, the telecommunications infrastructure requires power that is delivered by the power infrastructure. If power delivery is disrupted by disturbances in the power grid, telecommunication substations will run for a while on reserve battery power but would then need to switch to reserve backup generators (if they have them). The generator's operation would rely on fuel, first from on-site storage and then conveyed to a central distribution point by the energy distribution infrastructure and delivered to the telecommunications substation by the transportation infrastructure and paid for by the components of the financial infrastructure. The technicians who show up, through the transportation infrastructure, to make repairs would not do so unless they have been sustained by the food and water delivery infrastructures, and so forth. In turn, a functioning telecommunications system provides critical situational awareness and control to a power infrastructure that must keep its power generation in balance with its load in a dynamic control process over a very large geographical area. Telecommunications also plays a critical role in controlling the transportation system and is the basis of data exchange within the financial infrastructure. The complex interdependence between elements within each infrastructure is suggested and illustrated schematically but by no means wholly characterized by **figure 1-7**.

"Communicating across disciplines requires domain experts to learn one another's language to pose significant questions and usefully interpret answers," National Academy of Sciences, *Making the Nation Safer; The Role of Science and Technology in Countering Terrorism*

In the course of ordinary interruptions, many of these infrastructure interdependencies and interactions can be safely ignored. In an EMP attack scenario, the immediate insult is expected to affect the different infrastructures simultaneously through multiple electronic component disruptions and failures over a wide geographical area. Understanding these cross-cutting interdependencies and interactions is critical to assessing the capability of the full system of interdependent critical infrastructures to recover. The modeling and simulation needed to explore the response of such a complex situation involves a large but finite number of elements and should be amenable to analysis, at least approximately, but little effort has been made to address the problem to date.

In practice, understanding the interdependence may be a difficult task because subject area experts are not necessarily attuned to coupling mechanisms that span the boundary between their respective discipline and another, and because an accurate representation of the interdependence requires a familiarity with transdisciplinary phenomena.

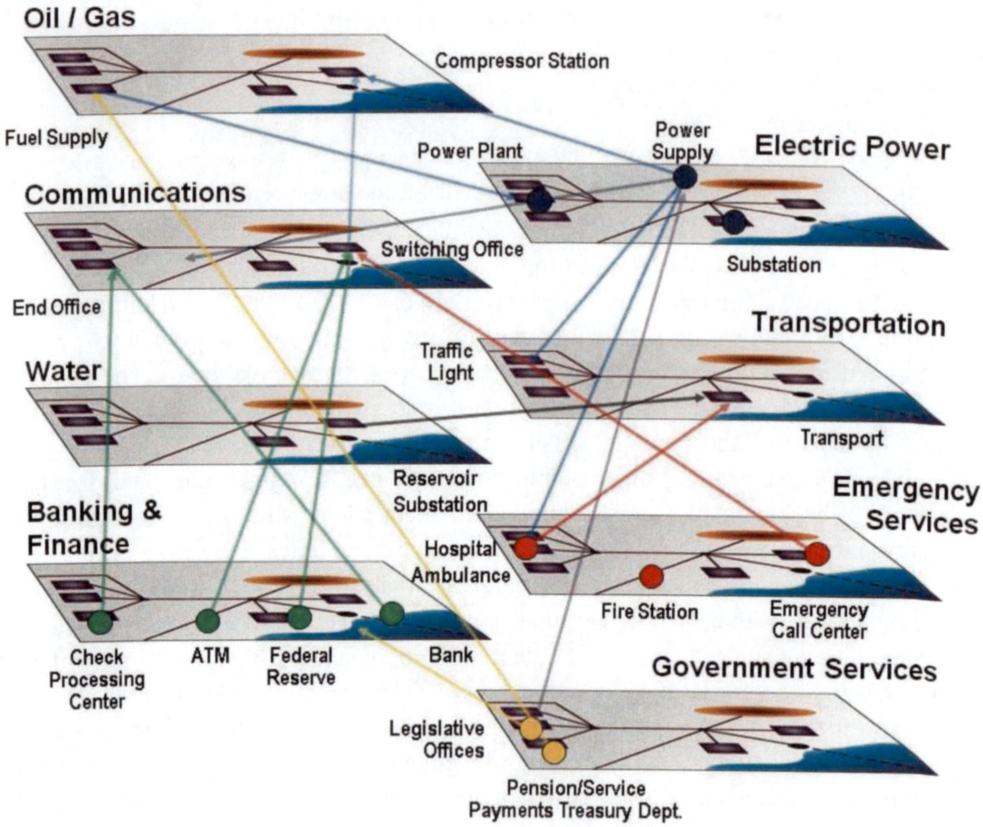

Some connections are not shown (diagram provided courtesy of Sandia National Laboratory).

Figure 1-7. A Conceptual Illustration of the Interconnectedness of Elements Contained Within Each Critical Infrastructure.

Experience demonstrates that it is sometimes easy to overlook the less obvious roles that such interdependencies and interactions may play, and coupling pathways may be easily overlooked. As an example, many of the recovery procedures developed by organizations to deal with emergencies involve the implicit assumption that transportation is available and people will be put on airplanes and go somewhere to diagnose and repair something. In the immediate aftermath of 9/11, all civilian airplanes were grounded. In 1991, a single point failure inside the telecommunications system, the accidental severing of a single fiber-optic cable in the New York City region, not only blocked 60 percent of all calls into and out of New York, but also disabled all air traffic control functions from Washington, D.C., to Boston — the busiest flight corridor in the Nation — and crippled the operations of the New York Mercantile Exchange.[2] These key interdependencies were always there, but they were not recognized as warranting advanced contingency planning, situational awareness in degraded conditions, and operational workarounds.

Recent Studies and Organizational Activities

Infrastructure vulnerability has been the subject of recent high-level attention with three separate congressionally chartered commissions devoted to the topic, including the President's Commission on Critical Infrastructures (the Marsh Commission), the EMP Commission and The National Research Council of the National Academy of Sciences. The latter issued a report, *Making the Nation Safer: The Role of Science and Technology in Countering Terrorism* in 2002, which explores potential vulnerabilities in the same list of critical infrastructures cited in the previous section. Both commissions noted the lack of a mature modeling and simulation capability as a significant weakness in the protective toolset available to planners and those charged with the mission of shielding our key infrastructures from subversion or other disruption. The National Academy of Sciences study, in particular, recommended the development of an analytic capability based on systems engineering principles.

An organizational response is beginning as well. For example, the Department of Homeland Security has absorbed and reorganized the National Infrastructures Analysis Center (NIAC), as well as the National Infrastructure Simulation and Analysis Center (NISAC). Many other government agencies and organizations have organized critical infrastructure protection organizations. Important infrastructure modeling work has been sponsored by the Government in such organizations as the Department of Energy's Argonne National Laboratory and National Nuclear Security Administration's (NNSA) laboratories at Sandia and Los Alamos.

Critical infrastructure studies also have been a growing activity in academia with the participation of individual scholars at various universities around the country. A number of academic centers have also set up or spun off entire institutes devoted to the analysis of critical infrastructural matters. The University of Virginia has created the Center for Risk Management, which focuses on the application of input-output econometric models to analysis of critical infrastructures. In addition, George Mason and James Madison universities in Virginia have created the Center for Infrastructure Protection Programs (CIPP). The Santa Fe Institute of Complexity Studies also has pursued important theoretical work, and there are many other examples as well.

These efforts, as well as other important related work, are pointing in the right direction. Nevertheless, the bottom-line is that currently an adequate capability to model individual infrastructures on a national scale does not exist. Moreover, the capability to develop and integrate a fully interactive and coupled set of national-scale infrastructure models is not being pursued with sufficient priority and support to achieve it in the foreseeable future.

Commission-Sponsored Modeling and Simulation (M&S) Activities

As the Commission embarked on its task, it attempted to engage existing capabilities within academia, industry, and government to simulate the behavior of infrastructures subjected to stressful disruption. To that end, it initiated the following activities.

National Workshop The EMP Commission sponsored a national workshop on the modeling and simulation of interdependent interacting infrastructures as part of an effort to understand the state of modeling capability in this country and to identify capabilities that might be exploited to provide insight into the expected effects of a prescribed EMP attack scenario. A number of national experts who are working on related modeling and simulation activities participated. The Commission has exploited some of these capabilities to develop insight that helped inform the assessment provided by the Commission's full report.

Contractual Activities Current modeling and simulation tools are not sufficient to provide a realistic predictive capability for the interdependent infrastructures. Nevertheless, the modeling capability proved useful in developing the Commission's insight into the effects of coupling on the overall impact due to the attack and the expected recovery and restoration effort. The Commission examined such questions as: Does a strong or weak coupling tend to drive the models to longer or shorter infrastructure restoration times? What seemed to be the sensitive parameters? What sorts of decoupling activities might be suggested to shorten reconstitution efforts? The examination of these and similar questions was supported by a number of efforts the Commission initiated with the NISAC, the University of Virginia, and Argonne National Laboratory. Some of the results of these efforts are summarized in the following section.

EMP Commission Staff Analyses The EMP Commission staff also developed analytic products to explore issues of stability and instability related to infrastructural coupling models. In particular, the Commission focused on models that coupled the power to the telecommunication infrastructure in an interactive way.

The results of these efforts have informed the Commission's findings, as documented in this volume.

Illustrative Modeling and Simulation Results for Coupled Infrastructures

To illustrate some of the complex behavior that can arise when coupling between infrastructures is included, consider the simple case of the interaction of only two infrastructures, here taken to be the telecommunication and power networks. The telecommunication networks themselves are in the midst of a rapid evolution that has seen data communications, which represented only 10 percent of the total traffic in 1990, grow to about 50 percent of the daily telecommunications load, with the expectation that voice traffic will represent only a small fraction of the traffic by 2015. There is a corresponding ongoing evolution, both in the network architecture and the underlying hardware, that is described in more detail in Section 3 of this report.

A critical element of the network of the future will be reliance on public data networks (PDNs). In the past, the electric power grid relied on its own communication system to monitor and control the grid, and mutual dependence between the power and telecommunications systems was essentially nil. Today the power grid relies on PDNs for about 15 percent of its telecommunication needs, and this figure is expected to grow to 50 percent in the near future. **Figure 1-8** illustrates the expected interdependence for this evolving network.

The PDNs represent networks powered by the power distribution network. The power generation and distribution network is, in turn, controlled by SCADA systems that depend on telecommunications to provide situational awareness and to execute control functions for the power grid. **Figure 1-9** represents the results of a model simulation. The telecommunication network reverts back to a dependence on commercial power while both are in the recovery phase. The power infrastructure continues to depend in part on the probability of call blocking, while the recovery for telecommunications depends on the available power.

The figure shows four distinct phases of a recovery process — an early phase extending to about a half hour, during which many network elements execute reinitializations to restore some service with power generally available from battery backup, to a phase of interdependency, during which the only power option left is reliance on the commercial grid, which in turn is dependent on a commercial PDN. This model can provide insight into the recovery process. It predicts significantly lengthened recovery times because of infrastructure interdependence, compared to recovery analysis that examines an infrastructure in isolation, ignoring the factor of interdependence. While illustrative of the effects of interdependency, this model is not meant to represent the actual behavior of any specific real-world system today.

In another effort, the NISAC studied the consequences of an EMP attack scenario[4] involving a large EMP source located at high altitude off the California coastline. The simulation looked at the effects on water, electric power, telecommunications, natural gas, refined petroleum products, transportation, labor and economic sector productivity and attempted to capture their known interactions. The simulation included network models for the transfer of information and infrastructure products, services, markets, and process models for each product and service. The boundary conditions for the simulation were provided by the EMP Commission; for study purposes, they included descriptions of the potential initial states of both the power and telecommunication infrastructures immediately following exposure to the EMP environment. The simulation, which was not considered realistic because it did not consider likely physical damage that would impede any recovery process, was still useful in providing insight into the potential for disturbances in one infrastructure to cascade into others.

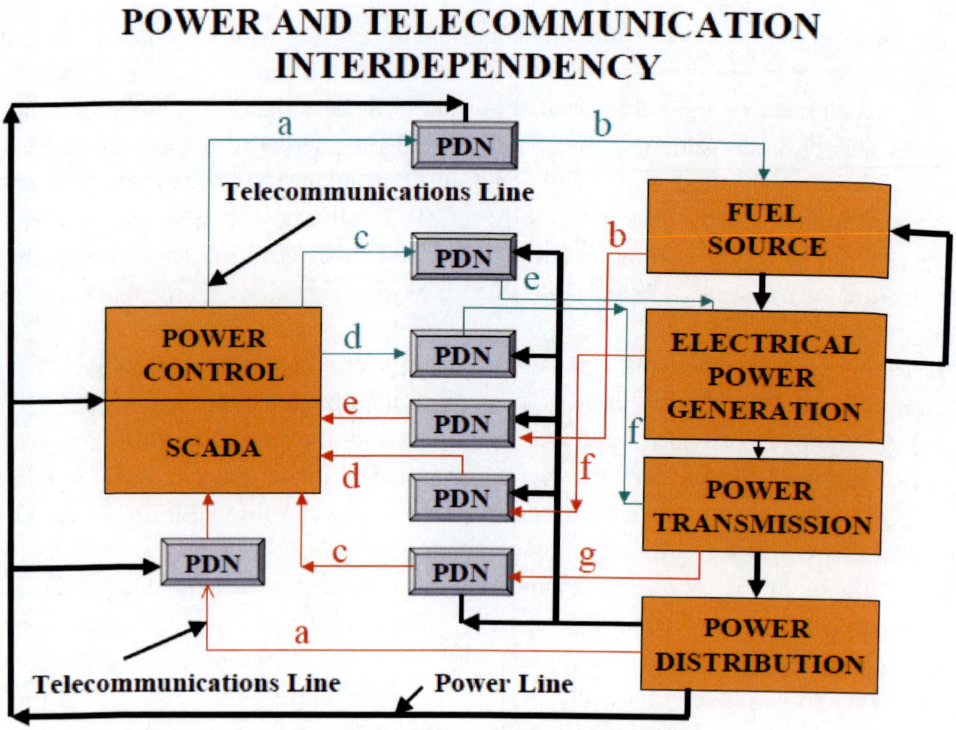

Figure 1-8. Interdependency for Anticipated Network of the Future.

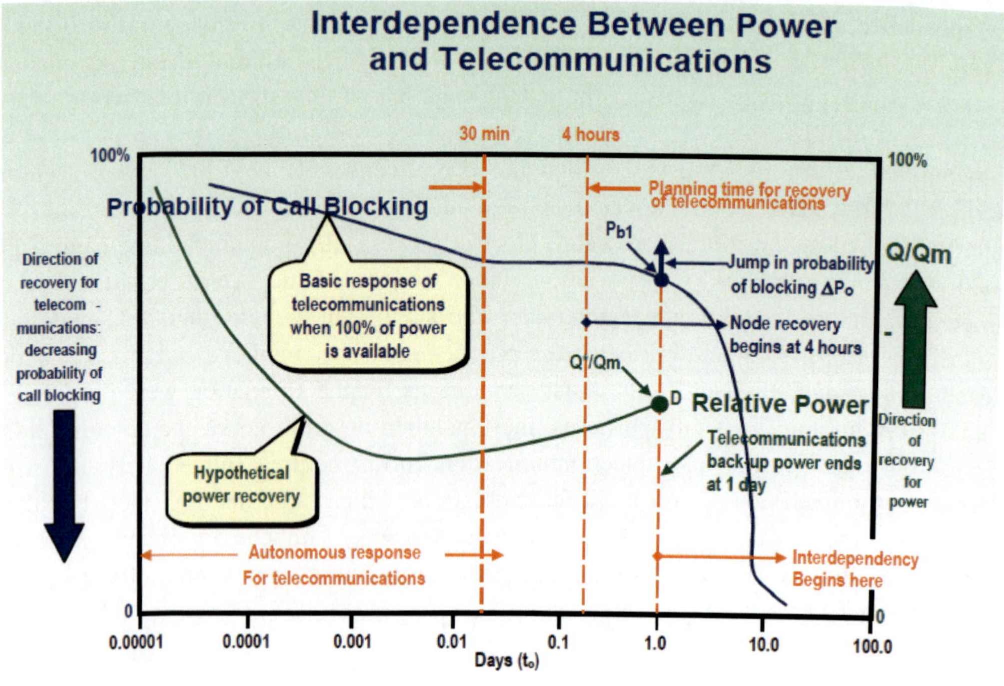

Figure 1-9. Results of a Model Simulation.[3]

Summary

No currently available modeling and simulation tools exist that can adequately address the consequences of disruptions and failures occurring simultaneously in different critical infrastructures that are dynamically interdependent. Many infrastructure models that do exist are local to regional in scope.

The Federal Government is supporting a number of initiatives to develop critical national infrastructure modeling and simulation capability as a national analysis and planning resource. However, these are not high national priorities and are funded at less than critical mass. They also are fragmented and uncoordinated, which is not an entirely negative observation, as the complexity of the task merits exploration of independent research and development approaches.

Recent analytic work suggests that evolving interdependencies may be inadvertently introducing entirely new and potentially serious vulnerabilities that could lead to infrastructure failures, even without the precipitating catalyst of an EMP attack.

Recommendations

- The Commission recommends that research be conducted to better understand infrastructure system interdependencies and interactions, along with the effects of various EMP attack scenarios. In particular, the Commission recommends that such research include a strong component of interdependency modeling. Funding could be directed through a number of avenues, including through the National Science Foundation and the Department of Homeland Security.
- The Commission recognizes current interest in protecting SCADA systems from electronic cyber assault. The Commission recommends that such activities be expanded to address the vulnerability of SCADA systems to other forms of electronic assault, such as EMP.

SECTION 2. ELECTRIC POWER

Introduction

The functioning of society and the economy is critically dependent upon the availability of electricity. Essentially every aspect of American society requires electrical power to function. Contemporary U.S. society is not structured, nor does it have the means, to provide for the needs of nearly 300 million Americans without electricity. Continued electrical supply is necessary for sustaining water supplies, production and distribution of food, fuel, communications, and everything else that is a part of our economy. Continuous, reliable electrical supply within very tight frequency boundaries is a critical element to the continued existence and growth of the United States and most developed countries.

For most Americans, production of goods and services and most of life's activities stop during a power outage. Not only is it impossible to perform many everyday domestic and workplace tasks, but also people must divert their time to dealing with the consequences of having no electricity. In the extreme, they must focus on survival itself. The situation is not different for the economy at large. No other infrastructure could, by its own collapse alone, create such an outcome. All other infrastructures rely on electric power. Conversely, the electric power infrastructure is dependent on other infrastructures that are themselves vulnerable to the direct effects of electromagnetic pulse (EMP) in ways that are described elsewhere in this report. No infrastructure other than electric power has the potential for nearly complete collapse in the event of a sufficiently robust EMP attack. While a less robust attack could result in less catastrophic outcomes, those outcomes would still have serious consequences and threaten national security.

The electrical power system is the largest single capital-intensive infrastructure in North America. It is an enormously complex system of systems containing fuel production, gathering and delivery systems, electrical generators (often themselves systems), electrical transmission systems, control systems of all types, and distribution systems right down to the electrical outlet and interconnection at the point of use. It is this vast array of systems and components all acting in concert, integrated into a cohesive whole to deliver electrical power at the point of use, with supply-on-demand at a uniform frequency that provides the reliable, steady, and adequate electric supply on which everyone has come to expect and depend. Because of the integration and interdependence of the electric system's components and the ever growing shift to electronics and particularly microelectronics for operation, protection and control, the Nation is particularly vulnerable to a major disruption of the electric supply.

Today, the existing electrical system at peak demand periods increasingly operates at or near reliability limits of its physical capacity. Modern electronics, communications, protection, control and computers have allowed the physical system to be utilized fully with ever smaller margins for error. Therefore, a relatively modest upset to the system can cause functional collapse. As the system grows in complexity and interdependence, restoration from collapse or loss of significant portions of the system becomes exceedingly difficult. Over the last decade or two, relatively few new large-capacity electric transmission capabilities have been constructed and most of the additions to generation capacity that have been made have been located considerable distances from load for environmental, political, and economic reasons, adding stress and further limiting the system's ability to withstand disruption. Significant elements of the system, including many generating plants, are aging (a considerable number are more than 50 years old) and becoming less reliable or are under pressure to be retired for environmental considerations, further exacerbating the situation.

Should the electrical power system be lost for any substantial period of time, the Commission believes that the consequences are likely to be catastrophic to civilian society. Machines will stop; transportation and communication will be severely restricted; heating, cooling, and lighting will cease; food and water supplies will be interrupted; and many people may die. "Substantial period" is not quantifiable but generally outages that last for a week or more and affect a very large geographic region without sufficient support from outside the outage area would qualify. EMP represents such a threat; it is one event that may couple ultimately unmanageable currents and voltages into an electrical system routinely operated with little margin and cause the collapse of large portions of the electrical system. In fact, the Commission is deeply concerned that such impacts are certain in an EMP event unless practical steps are taken to provide protection for critical elements of the electric system and to provide for rapid restoration of service, particularly to essential loads.

The electrical power system routinely experiences disruptions. In most cases, the cause is the failure of one or a small number of components. The overall system has a degree of durability against such failures, although in some cases failures lead to a cascading loss of power up to a regional level that extends over relatively short to moderate periods of time. The current strategy for recovering from such failures is based on the assumption of sporadic failures of small numbers of components, and for larger failures, drawing on resources from outside the affected area. This strategy leaves us ill-prepared to respond effectively to an EMP attack that would potentially result in damage to vast numbers of components nearly simultaneously over an unprecedented geographic scale.

The Commission recognizes that EMP is one of several threats to the overall electrical power system. Some of these threats are naturally occurring, such as geomagnetic storms. Others, like attacks using information operations on the system's controls, are manmade. There are strong similarities in the types of damage resulting from the occurrence of such threats. There are also similarities in the measures that are appropriate to be undertaken to reduce the electrical power system's vulnerability to each of these threats. The Commission believes that the measures it recommends will both reduce the vulnerability of the electrical power system to these threats and improve the Nation's ability to recover the system.

The magnitude of an EMP event varies with the type, design and yield of the weapon, as well as its placement. The Commission has concluded that even a relatively modest-to- small yield weapon of particular characteristics, using design and fabrication information already disseminated through licit and illicit means, can produce a potentially devastating E1 field strength over very large geographical regions. This followed by E2 impacts, and in some cases serious E3 impacts operating on electrical components left relatively unprotected by E1, can be extremely damaging. (E3 requires a greater yield to produce major effects.) Indeed, the Commission determined that such weapon devices not only could be readily built and delivered, but also the specifics of these devices have been illicitly trafficked for the past quarter-century. The field strengths of such weapons may be much higher than those used by the Commission for testing threshold failure levels of electrical system components and subsystems.

Additionally, analyses available from foreign sources suggest that amplitudes and frequency content of EMP fields from bomb blasts calculated by U.S. analysts may be too low. While this matter is a highly technical issue that awaits further investigation by U.S. scientific experts, it raises the specter of increased uncertainty about the adequacy of current U.S. EMP mitigation approaches.

A key issue for the Commission in assessing the impact of such a disruption to the Nation's electrical system was not only the unprecedented widespread nature of the outage (e.g., the cascading effects from even one or two relatively small weapons exploded in optimum location in space at present would almost certainly shut down an entire interconnected electrical power system, perhaps affecting as much as 70 percent or possibly more of the United States, all in an instant) but more significantly widespread damage may well adversely impact the time to recover and thus have a potentially catastrophic impact.

For highly dependent systems such as commercial telecommunications and the financial system, electric power is frequently filtered through batteries. These act to condition the power as well as to provide limited backup. Local, at-site emergency generators are used quite extensively for high priority loads. These include hospitals, cold storage, water systems, airport controls, rail controls and similar uses. These systems, however, are themselves increasingly dependent on electronics to initiate start up, segregate them from the larger power system, and control their operating efficiency, thereby rendering them vulnerable to EMP.

Furthermore, emergency generators have relatively short-term fuel supplies, generally less than 72 hours. Increasingly, locally stored fuel in buildings and cities is being reduced for fire safety (after 9/11) and environmental pollution reasons, so that emergency generation availability without refueling is becoming even more limited. Batteries normally have a useful life well short of emergency generators, often measured in a few hours. All of these tools for maintaining a stable and adequate power supply, even to high priority loads, are intended to be temporary at best – bridging the time until restoration can take place.

The impact of such an EMP-triggered outage would be severe but not catastrophic if the recovery was rapid or the geographic impact sufficiently limited. The recovery times from previous large-scale outages have been on the order of one to several days. This record of quick recovery is attributable to the remarkably effective operation of protective systems and communications that are an essential part of the power infrastructure and the multiple sources of replacement components from surrounding nonimpacted systems. In this context, a short blackout scenario over a relatively small geographic region would be economically painful. Of the more than $10 trillion U.S. Gross Domestic Product, about three percent is electricity sales. However, estimates of economic loss from historical blackouts range from factors of six (for domestic customers) to 20 (for industrial users) times the value of the interrupted service. By these measures, the economic impact of an outage is between 18 and 60 percent of total production in the affected area. Again, this estimate is for reasonably short-lived blackouts. A short blackout presents no threat to national survival.

On the other hand, a geographically widespread blackout that involves physical damage to thousands of components may produce a persistent outage that would far exceed historical experience, with potentially catastrophic effect. Simulation work sponsored by the Commission at the National Infrastructure Simulation and Analysis Center (NISAC) has suggested that, after a few days, what little production that does take place would be off-set by accumulating loss of perishables, collapse of businesses, loss of the financial systems and dislocation of the work force. The consequences of lack of food, heat (or air conditioning), water, waste disposal, medical, police, fire fighting support, and effective civil authority would threaten society itself.

The Commission solicited technical assistance and judgment from the North American Electric Reliability Corporation (NERC, which is governed by the Federal Energy Regulatory Commission [FERC] guidelines); utilities with particularly relevant experience (such as with geomagnetic storms [similar to E3]; long or very high voltage transmission; uniquely sensitive generation, special fault testing; and similar aspects); suppliers of protection, control, and other related equipment; groups dealing with industry standards; organizations of utilities, fuel suppliers, fuel transportation groups; select academic, national, and internationally recognized experts, the Department of Energy (DOE) National Laboratories, and relevant governmental entities. Willingness to be helpful was uniformly positive and generous. The Commission is grateful for this support.

NERC was uniquely well suited to be of assistance. NERC was established in the aftermath of the 1965 Northeast Power Failure to enhance the reliability of the electrical system. The Commission briefed the NERC Board of Trustees on the nature of the threat and the potential vulnerability. The NERC Board established an EMP task force under the aegis of its Critical Infrastructure Protection Advisory Group to provide technical advice to the Commission. The expertise of the task force membership spanned the three NERC Interconnects (Eastern, Western States Coordinating Council [WSSC], and Electric Reliability Council of Texas [ERCOT]), and all three major categories of the system (generation, transmission, and distribution).

This group's involvement was an essential element in focusing the Commission on the importance of the early-time EMP pulse and its implications for recovery, as well as on other triggers of widespread impact. It also provided technical input that was very helpful in implementing and interpreting a Commission-sponsored test program targeted at identifying the threshold at which significant control and protective components for the electrical system would begin to fail through disruption, false data, and damage. Many of the technical and operational insights discussed within the report were influenced by this task force although the NERC Task Force did not otherwise directly participate in the drafting of the report or in its conclusions.

Description

Major Elements

There are three major elements of the electrical power infrastructure: (1) generation, (2) transmission (relatively high voltage for long distances), and (3) distribution, whose elements are interdependent, yet distinct (see **figure 2-1**).

Generation

Power plants convert energy that is in some other form into electricity. The initial form of the energy can be mechanical (hydro, wind, or wave), chemical (hydrogen, coal, petroleum, refuse, natural gas, petroleum coke, or other solid combustible fuel), thermal (geothermal or solar), or nuclear. Power plants can range from single solar cells to huge central station complexes. In most circumstances the first stage of generation converts the original form of energy into rotational mechanical energy, as occurs in a turbine. The turbine then drives a generator.

Modern power plants all utilize complex protection and control systems to maximize efficiency and provide safety. They all have common electrical characteristics in order for them to be useable by all the various purposes to which electricity is put. Electronics have largely replaced all the electromechanical devices in older plants and are used exclusively in plants of the past one or two decades. Even generator exciters now have microprocessors and analog-to-digital converters. These electronics and, thus, the power plant itself are highly vulnerable to EMP assault. Identifying and locating damaged generation plant equipment with electronic sensors and communication interdicted and/or unreliable due to EMP and repairing the system would be a complex and time-consuming process, even when personnel and parts are readily available.

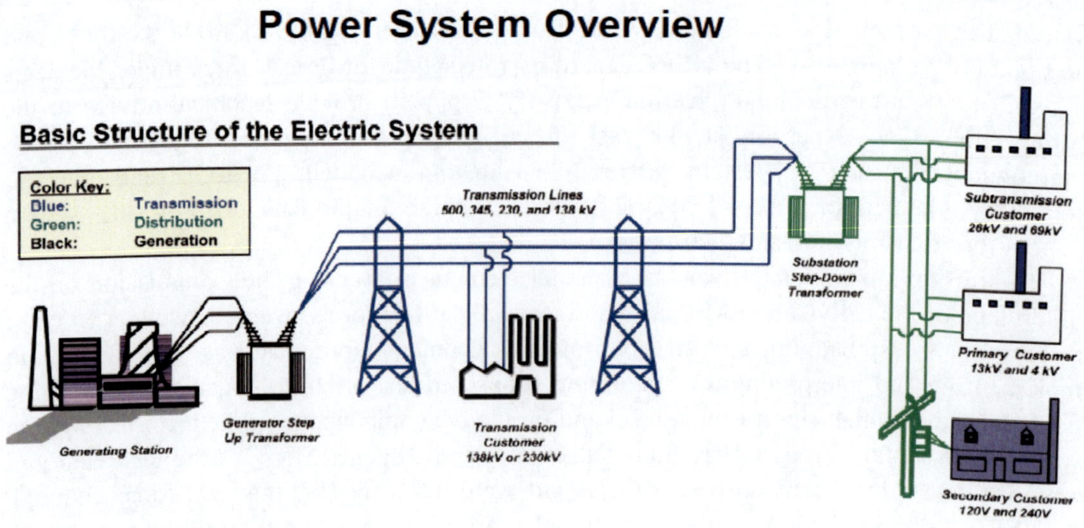

Figure 2-1. Power System Overview.

The fossil fuel supply system (coal, oil, wood, and natural gas) is largely dependent on electronics for its production and delivery of adequate fuel to the generators to produce nearly 75 percent of the Nation's electricity. There should not be a direct and immediate impact on the fuel supply for a nuclear power plant. The interdependency between the fuel necessary to generate electricity and the electricity and electronics to deliver the fuel is critical to the recovery. For example, natural gas normally is delivered just in time while oil and coal have some at-site storage. Nuclear generation supplies a major portion of the remainder of the Nation's electricity. It is unlikely for the timing of an EMP attack to be such that it would directly and immediately impact the fuel supply for a nuclear power plant. Of the balance, hydroelectric plants have their own fuel supplies as do geo-thermal, solar, and wind systems. However, wind and solar may or may not be generating in any event, given their inherent uncertainty. Hydro and geothermal are significant capabilities, but they are highly localized.

Transmission

Electrical power from the various power plants travels over a system of lines and substations to the regions and locales where it will be consumed. The transmission system moves large amounts of power generally over substantial distances and/or to directly serve very large electrical loads. This definition separates it from the distribution system, which is described below. Transmission includes lines (wires strung from insulator strings on towers or underground in special insulated containers) and substations (nodal points where several lines intersect and protection and control functions are implemented). Within substations there are transformers (which transform power from one voltage to another); breakers (similar to on-and-off switches able to handle the large amounts of energy passing through); and protective devices, meters, and data transmitting and control systems. Protective devices protect the electrical components from unusual electrical disturbances that occur from time to time for many different reasons as well as for general safety reasons.

The delivery of electrical power across or through some medium, such as a wire, encounters resistance, which itself takes power to overcome. Electrical power is measured by the product of voltage and current. The electrical resistive losses (restricting the flow) are proportional to the square of the current. Thus it is most efficient to transmit power at the minimum current that is practical (this results in the highest voltage for the same amount of power). Otherwise, more power is consumed just to push the electricity through or over a path with higher resistance.

Standard values for modern alternating current (AC) transmission line voltage range from 115 kV (115 thousand volts) to 765 kV, although some 1100 kV transmission has been developed and tested. The current carried by these lines is typically up to a few thousand amperes. Direct current (DC) is also used in some instances for moving large amounts of power great distances and for controlling the flow itself. The normal point of use of electricity is AC and thus the shift from AC to DC and back from DC to AC makes DC uneconomical other than in special circumstances. The use of DC is increasing, however, as power costs continue to grow and the technology to shift from AC to DC and back becomes less expensive. Transformers within the substations are used to move the voltage from one line or power plant up to or down to another voltage while maintaining essentially the same level of power.

Distribution

Loads or end users of electricity (residences, commercial establishments, and even most industry) require electrical power to be available in the voltages needed in adequate supply when they need it. This often means in relatively small quantities at low voltage and current. The size of the wires and switches in a typical house are able to be quite small and of much lower cost because the power available to that house is restricted to be relatively low. The electrical and electronic appliances similarly need only a small amount of power to be available. Therefore, the high-voltage power of the transmission system described previously is reduced (stepped down) through transformers and distributed to the end users in levels they need and can use. Reactive load balancing equipment is also part of the distribution system. This equipment is needed for system stability. The electrical power system's stability is finely tuned and fragile. Large-scale failures most often occur because the system is destabilized by local anomalies.

The distinction between transmission and distribution is sometimes a fuzzy one because it depends on the size and need of the load and the specific system involved. The distinction is relevant for regulatory and business purposes. It does vary somewhat from region to region. Traditionally distribution distances are under 20 miles and voltages are less than 69.5 kV (more commonly 13.5 kV). However voltages up to 115 kV are used in some locations. Distribution has substations just like transmission, only smaller. These are not manned. Of importance is that the local switching, controls, and critical equipment have become largely electronic with concomitant vulnerability to EMP.

Alternating current, as opposed to direct current, is the medium for use of electricity as a general matter. Electricity production, transmission, distribution, and use require a precise frequency. Thus it is necessary across the vast electrical power system to precisely and reliably synchronize the frequency and phase of power coming from different generating sources and reaching and being utilized by different loads. Testimony to the accuracy of this control has been the wide use and dependence on electric clocks and the functioning of many electronic devices. The difficulty of maintaining the frequency synchronization during off-normal conditions is usually a factor in large-scale power out-ages. For example, when the frequency moves very far from a constant required level, protective schemes at the generators within the transmission system and at the loads alarm and often automatically trip. Occasionally these trip out of proper sequence causing the system to compound rather than mitigate the problem, and the system collapses.

Control and Protection Systems

Overlaid on these three primary elements — generation, transmission and distribution — is a control system that directs the power where it is needed, maintains the frequency, and protects the system. Control is also necessary for commercial aspects. The controls must protect the system from transients such as lightning, correct synchronization errors by activating reactive sources or loads, isolate malfunctioning elements of the grid, and prevent self-damage from improper compensation or human error. The control systems also enable the deregulated energy marketplace by tracking the origin, route, and destination of the energy commodity. Central to the monitoring and coordination of the power grid is a broad class of devices called supervisory control and data acquisition (SCADA) systems. These conform to an agreed set of standards that make it possible to network many such systems over a generic communications system, regardless of modality. SCADA devices are in broad use in a variety of applications other than power.

The revolution in communication, information, system and component protection, and control technologies has reached essentially every segment of the economy, and its heavy impact on the electric power industry is no exception. The growing dependence of our infrastructures on ubiquitous electronic control and protection systems confers great benefits in terms of economic and operational efficiency, rapid diagnosis of problems, and real-time remote control. At the same time and less often remarked, it also represents a potential new vector of vulnerability that could be exploited by determined adversaries, and intellectual efforts to mitigate such threats have been engaged. The infrastructure's vulnerability to EMP and other broad-impact events raises the threat to an entirely new and vastly expanded plane of serious to catastrophic impacts.

Electronics have enabled electric power systems — generation, transmission, and distribution — to achieve greater levels of efficiency and safety with much lower adverse environmental impacts. Far less generation, transmission, and distribution are now neces- sary to provide the same amount of benefit to the end user, thus significantly enhancing productivity and overall quality of life. In doing so, however, the electrical system operates closer to theoretical capacity and thus at narrower margins of safety and reliability. Electronics have improved system economics and lowered the overall cost of power to the end user while reducing pressure on basic resources and limiting potential adverse impacts on the environment. This enhanced capability, both on the provider and consumer side, is in part responsible (along with the regulatory environment) for the low rate of investment in the high-value components of the electric system infrastructure. For example, slowly increasing electrical transmission demand has largely been met within the limits of current production capacity for these components.

The continuing evolution of electronic devices into systems that once were exclusively electromechanical, enabling computer control instead of direct human intervention and use of broad networks like the Internet, results in ever greater reliance on microelectronics and thus the present and sharply growing vulnerability of the power system to EMP attack. Just as the computer networks have opened the possibility to cyber assault on the power system or to electrical power system collapse associated with software failure (as during the August 14, 2003, blackout), they have provided an opportunistic pathway for EMP attack that is likely to be far more widespread, devastating, and difficult to assess and restore. Switches, relays, and even generator exciters now have microprocessors and analog-to-digital converters. These and other low-power electronics cannot be expected to withstand EMP-generated stresses unless they are well protected. Protection must encompass both device design and system integration. Even a well-designed system installed without regard for EMP intrusion via connecting lines can be rendered inoperative by EMP stress. There is a serious question regarding whether manual control of the system sufficient to allow continued service will be possible even at a much-reduced state in the aftermath of EMP.

The key vulnerable electronic systems are SCADA along with digital control systems (DCS) and programmable logic controllers (PLC). SCADAs are used for data acquisition and control over large and geographically distributed infrastructure systems while DCSs and PLCs are used in localized applications. These systems all share similar electronic components, generally representative of components that form the internal physical architectures of portable computers. The different acronyms by which we presently identify SCADA, DCS, and PLC should not obscure the fact that the electronics have evolved to the point where the differing taxonomies are more representative of the functional differences of the electronics equipment rather than differences in the electronics hardware itself.

Electronic control equipment and innovative use of electronic controllers in equipment that is not usually considered control equipment are rapidly replacing the purely electro-mechanical systems and devices that were their predecessors. The use of such control equipment is growing worldwide, and existing users are upgrading equipment as new functionalities develop. The U.S. power industry alone is investing about $1.4 billion annually in new SCADA equipment. This is perhaps 50 times the reinvestment rate in transformers for transmission. The present rate represents upgrade and replacement of the protection and control systems to ever more sophisticated microelectronics at roughly 25 to 30 percent annually, with each new component more susceptible to EMP than its predecessor. The shift to greater electronic controls, computers, and the Internet also results in fewer operators and different operator training. Thus the ability to operate the system in the absence of such electronics and computer-driven actions is fast disappearing. This is almost certain to have a highly deleterious effect on restoring service in the event of an EMP attack.

Electrical System Organization

The integrated electrical power system of the United States and integrated systems in Canada and Mexico are covered by the NERC. This vast network is broken into only three truly separate systems at the present time — the Eastern Interconnection, the Western Interconnection, and Texas. The dividing line geographically between the Eastern and Western systems is roughly a line between Montana and North Dakota continuing southward. The largest of these, the Eastern Interconnection, serves roughly 70 percent of the electrical load and population of the United States. The three regions are separated electrically in AC in order to provide barriers for transfer of major frequency deviations associated with system separations. This mode of operation between regions is referred to as maintaining frequency independence. Importantly, this also acts as a barrier to EMP-caused system disruption or any other major system disruption and consequent collapse crossing between these three regions.

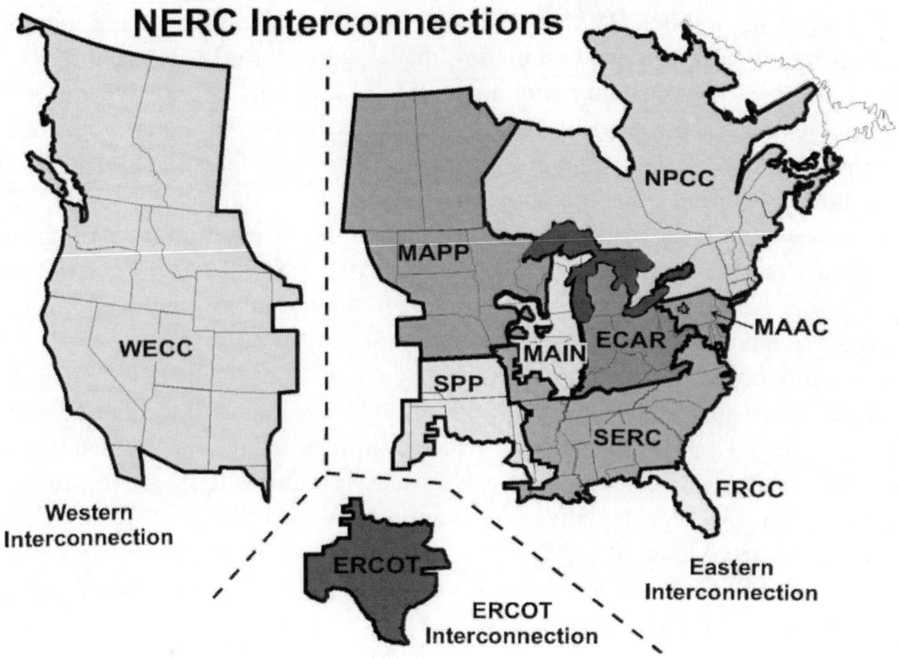

Figure 2-2. NERC Interconnections.

In **figure 2-2** the map of the three NERC regions shows the divisions geographically and the barriers for transfer of major frequency deviations associated with system separations. There are some nonsynchronous connections, such as DC back-to-back converter installations that facilitate limited power transfers yet maintain a barrier. The subregions identified in the map within a region are for organizational, record keeping, and management only. They do not have frequency independence from one another at this time. Thus at present, whole regions can be caused to collapse by sufficiently large electrical disturbances, like EMP, which severely exacerbates the problem of service to critical loads and importantly impedes restoration where delay increases the adverse impacts virtually exponentially.

Capacity Reserves

Although greater conservation and efficiency at the end user has reduced the need for new generation largely through the use of improved electronics and controls, the growing economy and use of ever greater labor- and material-saving devices continues to drive the need for new generation. Furthermore, older generation is being replaced for economic, environmental, and locational reasons. Increasing capital costs emanating from world market competition and natural disasters, plus the increasing cost of capital, have slowed the addition of new generation capacity. The inability in many cases to get generation to market with reasonable assurance due to limited transmission has similarly limited new generation additions. Finally, regulatory returns and pressure from competing uses of capital within utility systems or their parents, including municipal and public systems, have further restricted new generation of consequence. As a result, generation capacity margins have decreased.

Changes in the regulatory environment with greater deregulation of the generating sector have further encouraged recent increases in new generation capacity along with retirement of older units. Most of the new power plants over the past decade or two have been natural gas-fired units that are agile in their ability to adapt to market demands and opportunities, are relatively clean environmentally for fossil plants, faster to build and have lower capital cost than many alternative generator options. They have been located farther from load in most instances than the older plants or previously planned additions, and they are operated and integrated very differently than in the past as economic decisions are often driven by very diverse and nonintegrated responsibility. This can stretch the ability of the transmission system to get the new generation to load. The type and location of new generation stresses the system and increases its vulnerability to various threats including EMP.

The capacity margin (standby capacity for emergencies or other unplanned needs) for the transmission system grid (system of higher voltage lines and substations) has decreased from about 20 percent twenty years ago to about 10 percent now as an overall system matter although there are considerable regional or local variations. This reduced margin is due to little new construction, improved efficiency of the existing system, and the location of new generation away from load. It is further exacerbated by the addition of significant generation from renewable resources such as wind energy, which operates when the wind blows, not when the electrical system might otherwise require power. This results in shifting the generation between the wind and other controllable generation on an unpredictable basis regardless of the transmission system reliability needs, all of which results in greater and less predictable stresses on the overall system.

Operation of the transmission system at today's reduced margin while maintaining excellent reliability has been enabled by improved technology and operating practices for protection, command, and control of the transmission grid. While power production and consumption have grown, almost all of the growth has been absorbed on existing power lines although new substations have been added. There has been very little construction of transmission capacity, particularly of new longer distance transmission lines, or renewal and replacement of existing infrastructure for many reasons, including deregulation (discussed in the next section of this Section). The transmission system thus is operating with little ability to absorb adverse electrical impacts.

Overall, as a result of reduced generation capacity margins, the generation component of the system is far less able to compensate for the difficulties that may be encountered within the transmission system and vice versa. Together, the consequence is a power system far more vulnerable to disruption than in the past, and this vulnerability is increasing. While greater protection and control schemes have still provided a very reliable system in spite of this, the system is being stressed beyond reasonable limits. The electrical power system has become virtually fully dependent upon electronic systems working nearly flawlessly. The overall system reliability is testimony to the skill and effectiveness of the control systems. However, the lack of margin (combination of generation and transmission margins) results in making catastrophic cascading outages far more likely, and should the electronics be disrupted, the system is highly likely to fail on a broad scale. Thus, the small margin and reliance on electronics give rise to EMP vulnerability.

High-value assets (assets that are critical to the production and delivery of large volumes of electrical power and those critical for service to key loads) in the system are vulnerable to EMP through the loss of protection equipment due to E1 and even if E3 levels were not large enough to cause damage. The largest and most critical of these are transformers. Transformers are the critical link (1) between generation and transmission, (2) within the transmission network, (3) between the transmission and distribution systems, and (4) from the distribution to the load.

The transformers that handle electrical power within the transmission system and its interfaces with the generation and distribution systems are large, expensive, and to a considerable extent, custom built. The transmission system is far less standardized than the power plants are, which themselves are somewhat unique from one to another. All production for these large transformers used in the United States is currently offshore. Delivery time for these items under benign circumstances is typically one to two years. There are about 2,000 such transformers rated at or above 345 kV in the United States with about 1 percent per year being replaced due to failure or by the addition of new ones. Worldwide production capacity is less than 100 units per year and serves a world market, one that is growing at a rapid rate in such countries as China and India. Delivery of a new large transformer ordered today is nearly 3 years, including both manufacturing and transportation. An event damaging several of these transformers at once means it may extend the delivery times to well beyond current time frames as production is taxed. The resulting impact on timing for restoration can be devastating. Lack of high voltage equipment manufacturing capacity represents a glaring weakness in our survival and recovery to the extent these transformers are vulnerable. Distribution capability is roughly in the same condition although current delivery times are much less (i.e., limited manufacturing capability, although there is domestic production).

Deregulation

At least a decade ago, the power systems were owned and operated by vertically integrated utility companies. These entities consisted of investor-owned (owned by share-holders, commonly referred to as private) utilities, utilities that are government constructs (federal, such as the Tennessee Valley Authority, Bonneville Power Administration, and others), consumer-owned cooperatives, municipalities, and entities of the state such as peoples and public utility districts. The different entities were granted monopoly powers for service and were regulated through a variety of mechanisms including self-regulation for some of the government entities. In any given service territory, the local utility owned the generation, transmission, and distribution and was responsible for adequate supply, reliability, and other aspects of service quality.

This situation has changed. On April 24, 1996, FERC issued Orders 888 and 889, which encouraged wholesale power supply competition, deregulating this single aspect of the electrical industry. This allowed any party to produce and sell power to any other party at the wholesale levels (meaning sales to utility or load-serving entities, as opposed to direct retail sales to end users). Existing generation by investor-owned utilities was forced to be divested in many circumstances. This regulation applied only to the investor-owned utilities comprising a bit more than half the total U.S. electrical load. Many governmental or public entities did not possess generation of their own, and some that did followed by example and market imperatives. In some instances states have carried the deregulation further to a variety of forms of competition at the retail level.

The transmission infrastructure has remained regulated, and the previous vertically integrated systems in many instances were not allowed to commercially control their transmission in order to free up competition at the wholesale levels. Due to the complexity of an open market using an infrastructure that was built for another operating environment, the requirements for investment in the transmission system have been uncertain and more expensive. FERC regulates the transmission facilities in terms of use and pricing, but not location. The federal transmission regulation paradigm is moving toward being market based, which will have unknown impacts but is believed to assist in the development of new transmission facilities. The states also play an important role in the regulation of transmission and generation that is not consistent from state to state. With the market (and the market model itself) in flux, there is unwillingness presently to invest in transmission infrastructure.

There is no incentive for the states or localities to accede to construction of lines that are to move power over or through the state or locality without direct benefit to such state or locality. The power going through the lines pays no fees and no taxes to the hosts, although there are minor property taxes on the physical facilities in some instances. Until recently there was no capability to track the path of a given unit of energy when operating in AC and even then it is more calculation via model-specific than actual measurement. This is because AC power travels over the path of least resistance not as the flow of power might be contracted. Thus while a new interconnected line may appear to carry power pursuant to a contract for delivery between two parties, it is unlikely that the power will flow physically as envisioned. Thus it is unclear who will pay for the use of the new line. While new capability to track AC power (E-tags) could provide the basis for fiscal incentives for new line construction, it is not yet widely deployed nor well understood or accepted. Moreover, regulatory requirements create impediments to new line construction even if the incentives and capital are at hand. In short, from a business perspective, transmission lines are often low return or loss centers in the current environment.

The end state of the regulatory paradigm is still undetermined, and this uncertainty coupled with lack of local benefits when passing through state and local areas all contribute to the diminishing transmission capacity margins. There is uncertainty whether, by the time construction of new lines is completed, the investment could be recovered. It is likely that this situation will persist until the market model is clarified and implemented, which may take several years given the complexity and number of competing interests, including between the states and the Federal Government as well as with neighboring states. The market and system reliability pressure may move this faster as recognition and evidence mount. As noted earlier, the reduced and diminishing margins contribute significantly to EMP vulnerability.

Vulnerabilities

In order to assess the nature of EMP effects on the electrical system, we separately analyzed the potential effects of an electromagnetic pulse on each of the three main constituents of the power system — generation, transmission, and distribution. Within the context of a principal finding of the NERC EMP task force, recovery following an EMP-caused outage within any reasonably acceptable time is contingent largely on preventing damage to the high-value assets (assets that are critical to the production and delivery of large volumes of electrical power and those critical for service to key loads) and identifying and replacing ones that become damaged. Therefore, the Commission focused on identifying what those high value assets might be and their susceptibility to EMP damage. Thus, proper design, installation, and functioning of the protective equipment for these assets during an EMP attack are critical. There are other critical aspects to recovery that are discussed subsequently.

Generation

A power plant is designed to protect itself in the event of instantaneous loss of load, electrical faults or trips on the interconnected transmission system or internally, frequency excursions beyond rather tight limits, and often for the loss of an external power source for proper shutdown. None of these conditions should damage a power plant if the protective systems function properly, as frequently has been demonstrated. Very little damage to generation has occurred in previous blackouts, including the August 14, 2003, blackout. However, some malfunctioning in the multiple controls throughout a power plant does occur, albeit rarely. Therefore, on a broad enough scale, as in an EMP attack affecting many power plants at once, damage to a small number of these power plants would be expected statistically. Since E2 and E3 are not assessed as direct threats to the generation system (except for their step-up transformers and associated breakers), the critical vulnerability question is E1-induced plant control system failure.

The E1 pulse can upset the protection and control system, including damaging control and protective system components, and cause the plant to trip or trigger emergency controlled shut down. Current, temperature, pressure, frequency, and other physical parameters are monitored by the control systems. These provide independent measurements of same system, and all can cause the plant to trip off line and go to controlled shut down. Given the redundancy of protective system design, either several protective devices or devices in the critical path would have to fail in order for the plant not to initiate protective shutdown. However, if the control system itself or secondary nodal controls and receivers critical to orderly shut down are themselves damaged, as is reasonably possible with E1, then the plant is seriously at risk. Power plants, particularly newer ones, are highly sophisticated, very high-speed machines, and improper shut down can damage or destroy any of the many critical components and can even cause a catastrophic failure. Nuclear plants are an exception due to the nature of their protection schemes.

Given the range of potential E1 levels, analysis and test results provide a basis to expect sufficient upset to cause a plant's system to shut down improperly in many cases. Proper shutdown depends on synchronized operation of multiple controllers and switches. For example: coal intake and exhaust turbines must operate together or else explosion or implosion of the furnace may occur. Cooling systems must respond properly to temperature changes during shut down or thermal gradients can cause boiler deformation or rupture. Orderly spin-down of the turbine is required to avoid shaft sagging and blades impacting the casings. Bearings can easily fail and freeze or damage the shaft if the shut down does not engage emergency lubrication. There are similar issues inside very complex machines operating at high temperatures at fast speeds with tight tolerances. Thus, power plant survivability depends on a great many protective systems creating multiple pathways to plant damage and failure.

Restoration of some damage can be very long term, certainly months and in some instances years. The loss of generation of any size itself would contribute to systemwide collapse and certainly would limit restoration. Manufacturers of generation plant protection and control equipment performed some limited evaluation and while there are layers of redundancy, as noted, more and more these systems are going to computer-controlled microelectronics, and thus are more susceptible to EMP disruption.

At the device level, power plant protective systems are less exposed than the corresponding systems in the transmission grid. They act on local information, so failure of telecommunications systems is not as much of an issue for plant protection where operators are available in most instances 24/7 and can independently assess the situation and act. The control equipment, protective systems, sensors, and current transformers typically (but by no means always) will be inside the plant although this does not necessarily mean they will not be exposed. In general there will be no outside cable runs, so the building itself will provide some EMP protection. However the lengths of these interior cables can be on the order of 100 meters. Cable trays may or may not provide additional protection, depending on their material and installation method. The key is not device- or component-level testing for EMP susceptibility but overall control and protective system test to evaluate vulnerability. Subjecting an entire sophisticated and modern power plant to testing is not feasible. However, it does not take many damaged plants out of the many hundreds to seriously impact the system operation and the ability to restore service. The fact that all power plants exposed to E1 EMP will be illuminated simultaneously (within one power cycle) makes the situation extremely serious.

System Restoration — Generation

The restoration of the system from collapse is very complex in operation, almost an art rather than a science, and it requires highly trained and experienced operators with considerable information and controls at hand. Basically, in isolated cases or when beginning restoration, a load and generation source has to be identified and interconnected without interference from other loads or generation. These are then matched and gradually restored together. Thereafter, each increment of generation and load is added in turn to a larger operating system of generation and load. As each component of load and generation are included, the frequency will be impacted. If it varies outside very tight limits, it will all trip off and have to be put back together again. In most system disruptions leading to blackouts, there are large amounts of system still intact on the periphery of the disruption, which are able to greatly assist in the restoration, more easily allowing and absorbing each addition of generation and load until all is restored.

Every generator requires a load to match its electrical output as every load requires electricity. In the case of the generator, it needs load so it does not over spin and fail, yet not so much load it cannot function. In a large integrated system, where increments of load and generation are not sufficient to cause the frequency to drop or rise above acceptable margins, it is relatively straightforward and commonplace, just as turning on a light switch causes a generator someplace to pick up the load. In the case where the system is being restored and there are few loads and generators connected, this matching requires careful management and communication between load and generation.

Generation start-up for most plants requires power from another source to drive pumps, fans, safety systems, fuel delivery, and so on. Some, like hydroelectric and smaller diesels can start directly or from battery sources assuming they can control their access to matching load. In the case of EMP, large geographic areas of the electrical system will be down, and there may be no existing system operating on the periphery for the generation and loads to be incrementally added with ease. Furthermore, recovery of lost generation would be impacted by the loss of other infrastructure in varying degrees according to the type of plant. In that instance, it is necessary to have a "black start": a start without external power source. Coal plants, nuclear plants, large gas- and oil-fired plants, geothermal plants, and some others all require power from another source to restart. In general, nuclear plants are not allowed to restart until and unless there are independent sources of power from the interconnected transmission grid to provide for independent shutdown power. This is a regulatory requirement for protection rather than a physical impediment. What might be the case in an emergency situation is for the Government to decide at the time.

Black-start generation is that kind of generator that is independent of outside power sources to get started, hence the term black start. Most black start units today are hydroelectric plants, small gas peaking units, small oil-fired peaking units and diesel units. In some cases the black start unit may be collocated with a larger power plant in order to get the larger one started for system restoration. Fuel supply would then be the only issue from the generation perspective; for example, a gas plant might not have the fuel due to EMP damage someplace in the delivery system. Assuming the black start units were not damaged by EMP or have been repaired and assuming they are large enough to be significant, workers can begin the system restoration as building blocks from the generation side of the equation. E1 may have also damaged their startup electronics, which will need to be repaired first. It is often the case that generation capable of black start is not manned, so if they fail to start remotely, a person will need to be dispatched to find the problem, locate the needed parts, and get it operating. There are not many black start-capable units in locations that are suitable to independent restoration at this time. Recovery in most regions therefore needs to wait for other areas to restore power and then be reconnected increment by increment.

Even if partially disabled control systems successfully protect the critical generating equipment, all affected plants would face a long process of testing and repairing control, protective, and sensor systems. Protective and safety systems have to be carefully checked out before start up or greater loss might occur. Repair of furnaces, boilers, turbines, blades, bearings, and other heavy high-value and long lead-time equipment would be limited by production and transportation availability once at-site spares are exhausted. While some spare components are at each site and sometimes in spare parts pools domestically, these would not cover very large high-value items in most cases, so external sources would be needed. Often supply from an external source can take many weeks or several months in the best of times, if only one plant is seeking repair, and sometimes a year or more. With multiple plants affected at the same time, let alone considering infrastructure impediments, restoration time would certainly become protracted.

Transmission

Most generation is located outside major population areas and thus sometimes at great distances from the load being served. In general, electricity often travels great distances on an efficient high-voltage transmission system. The transmission system is made up of different owners, voltage levels, and controls. Yet power must be routed to where it is needed, so there are nodes called substations where the power lines join and are switched, and where power is moved from one voltage level to another level, interconnected with other transmission system components, and sent on to distribution systems. Finally as it gets closer to load, power is stepped down (reduced in voltage) and then down again and often down yet again to and within the distribution system and then normally down again to the delivery point for the load. Each of those step-down points requires a transformer to effect the change and breakers to isolate the transformer when necessary.

In the event of the loss of a generation facility, a fully functional transmission system can move the remaining generation from whatever plants can operate to areas otherwise affected by loss of a particular generating station. This occurs in normal practice as generation plants are brought in and out of service for one reason or another. The same thing happens when part of the transmission system is down for whatever reason. Other transmission in the network picks up the loss and generation is shifted so that the loads can continue to be served. All this is accomplished regularly as part of system operation. The ability to adjust quickly given access to a multitude of resources, generation, and transmission makes the system reliable. Incapacitation of sufficient elements of the transmission system would mean the inability to deliver power whether the generation is available or not. The same inability would be true for incapacitation of sufficient generation. In the case of EMP, both would be likely to be impacted simultaneously. This is what results in a blackout where the load does not get served. The transmission system is highly vulnerable to EMP.

Substation control systems at the nodes or hubs in the transmission system are inherently more exposed to the E1 pulse than their power plant counterparts, which are often not in buildings at all. The sensors, communications, and power connections are outdoors and cables (i.e., antennas in the sense of an EMP receptor) which may be hundreds of meters long may be buried, run along the ground, or elevated. The control devices themselves, including the protective relays, may even be in remote structures that provide little electromagnetic attenuation. Most substations do not have operators present but are remotely controlled from power dispatch centers, in some instances hundreds of miles away.

Operation of transmission substations depends on various communications modalities, including telephone, microwave, power line communications, cell phones, satellite phones, the Internet, and others. Typically, these modes are used for dedicated purposes; they do not necessarily provide a multiple redundant system but are "stove piped." From the point of view of managing routine system perturbations and preventing their propagation, NERC advises us that the telephone remains the most important mode. If the voice communications were completely interrupted, it would be difficult, but still reasonably possible, to successfully continue operations — provided there were no significant system disruptions. However in the case of an EMP event with multiple simultaneous disruptions, continued operation is not possible. Restoration without some form of communication is also not possible. Communication is clearly critical in the path to restoration.

Just as in the case involving power plants, the first critical issue is the proper functioning of the protective elements, specifically relays, followed by the local control systems. These elements protect the high-voltage breakers and transformers that are high-value assets. High-value assets are those that are critical to system functioning and take a very long time to replace or repair. Other protected devices, such as capacitors and reactive power generators, are also high value and nearly as critical as the transformers. E1 is likely to disrupt and perhaps damage protective relays, not uniformly but in statistically very significant numbers. Left unprotected, as would likely result from E1 damage or degradation to the protective relays, the high-value assets would likely suffer damage by the transient currents produced during the system collapse, as well as potentially from E2 and E3 depending upon relative magnitudes. Commission testing of some typical protective relays with lower than expected EMP levels provides cause for serious concern.

The high-value transmission equipment is subject to potentially large stress from the E3 pulse. The E3 pulse is not a freely propagating wave like E1 and E2, but the result of distortions in the Earth's magnetic field caused by the upper atmosphere nuclear explosion. The distortion couples very efficiently to long transmission lines and induces quasi-direct current electrical currents to flow. The currents in these long lines can aggregate to become very large (minute-long ground-induced currents [GIC] of hundreds to thousands of amperes) sufficient to damage major electrical power system components. With respect to transformers, probably the hardest to replace quickly, this quasi-direct current, carried by all three phases on the primary windings of the transformer, drives the transformer to saturation, creating harmonics and reactive power. The harmonics cause transformer case heating and over-currents in capacitors potentially resulting in fires. The reactive power flow would add to the stresses on the grid if it were not already in a state of collapse. Historically, we know that geomagnetic storms, which can induce GIC flows similar to but less intense than those likely to be produced by E3, have caused transformer and capacitor damage even on properly protected equipment (see **figure 2-3**). Damage would be highly likely on equipment unprotected or partially protected due to E1.

The likelihood and scope of the E3 problem are exacerbated by the small transmission margins currently available. The closer a transformer is operating to its performance limit, the smaller the GIC needed to cause failure. Moreover, newer transmission substations are increasingly using three single-phase transformers to handle higher power transfer, since the equivalently rated three-phase transformers are too large to ship. The three- phase systems are more resistant to GIC, since their design presumes a balanced three- phase operation. Thus the separate single-phase transformers are more susceptible to damage from GIC.

System Restoration — Transmission

The transmission system is the lynch pin between generation and load. It is also a network interconnecting numerous individual loads and generating sources. To restore the overall power system to get generation to load, as noted earlier, an increment of generation needs to be matched to an increment of load and then add the next matching increments and so on. As the number of increments becomes greater, there is some flex in the system to absorb variations. As a result, the restoration is easier and goes much faster. In the initial increments however, the transmission system link between generation and load has to be isolated so other loads, which may well remain connected, do not impact the effort. This is tricky and requires careful coordination to adjust the breakers in the sub- stations so the link is routed correctly and safely.

The power transmission grid is designed to break into islands of hopefully matched generation and load when the system receives a sufficient electrical disruption. This is both to protect service in the nonimpacted regions and to allow for the stable systems to be used to restart the island that lost functionality. With EMP, broad geographic reach and simultaneous multiple levels of disruption result in a situation in which the islanding schemes themselves will probably fail to work in the EMP-affected area. Since the geographic area is so large, perhaps encompassing an entire NERC region or possibly more, restoring the system from the still functioning perimeter may well not be possible at all or would take a great deal of time; the Commission estimates weeks to months, at least in the best circumstance.

Figure 2-3. GIC Damage to Transformer During 1989 Geomagnetic Storm.

Distribution

Most of the long power outages that Americans have experienced were due to physical damage to the distribution system — local damage. This damage is usually caused by natural events such as weather. Windblown trees fall on neighborhood power lines or ice buildup drops lines that in some instances make contact with live lines causing arcs that in turn can even result in distribution transformers exploding.

EMP damage to the distribution system would be less dramatic than that inflicted upon the transmission system but still would result in loss of load. The principal effect of EMP would be E1-induced arcing across the insulators that separate the power lines from the supporting wood or metal poles. The arcing can damage the insulator itself and in some cases result in pole-mounted transformer explosions. Damage to large numbers of insulators and pole-mounted transformers could also result in a shortage of replacement parts, as these items are fairly reliable under normal conditions, and spares are not kept to cover widespread losses. Ultimately workarounds and replacements can be found in most circumstances although widespread damage and impact to related infrastructures will cause delay.

The important effect of the loss of load in the EMP scenario is that it happens simultaneously. Thus it represents a substantial upset to the entire grid, causing the frequency to spin up and protective relays to open on generation and can by itself result in a cascading failure and blackout of the entire NERC region. Similarly, any consumer or industrial electrical device that is shut down or damaged by EMP contributes to the load loss and further drives the system to collapse. It becomes a case of what comes first to cause what failure since the EMP E1 impulse is virtually simultaneously disrupting all facets of the electrical system and load.

Synergistic Effects of E1, E2, and E3

The effects of EMP on the electrical power system are fundamentally partitioned into its early, middle, and late time effects (caused by the E1, E2, and E3 components, respectively). The net impact on the electric power grid includes the synergistic interaction of all three, occurring nearly simultaneously over a large geographic area. The Commission has concluded that the electrical system within the NERC region so disrupted will collapse with near certainty. Thus one or more of the three integrated, frequency-independent NERC regions will be without electrical service. This loss is very large geographically and restoration is very likely to be beyond short-term emergency backup generators and batteries. Any reasonable EMP event would be much larger than the Texas region so basically the concern is the Eastern and Western regions with Texas either included or not depending upon the location of the weapon. The basic threat to U.S. society that moves an EMP event from a local or short-term adverse impact to a more prolonged and injurious event is the time it takes to restore electrical and other infrastructure service.

The early time EMP, or E1, is a freely propagating field with a rise time in the range of less than one to a few nanoseconds. E1 damages or disrupts electronics such as the SCADA, DCS, and PLC as well as communications and to some extent transportation (necessary for supplies and personnel). This disrupts control systems, sensors, communication systems, protective systems, generator systems, fuel systems, environmental mitigation systems and their related computers, as well as the ability to repair. SCADA components, in particular, are frequently situated in remote environments and operate without proximate human intervention. While their critical electronic elements are usually contained within some sort of metallic box, the enclosures' service as a protective Faraday cage is inadequate. Such metallic containers are designed only to provide protection from the weather and a modicum of physical security. They are not designed to protect the electronics from high-energy electromagnetic pulses, which may infiltrate either from the free field or from the many antennae (cable connections) that compromise electromagnetic integrity.

The E1 pulse also causes flashovers in the lower voltage distribution system, resulting in immediate broad geographic scale loss of electrical load and requiring line or insulator replacement for restoration.

The intermediate time EMP, or E2, is similar in frequency regime to lightning, but vastly more widespread, like thousands to millions of simultaneous lightning strikes, even if each strike is at lower amplitude than most naturally occurring lightning. The electrical power system has existing protective measures for lightning, which are probably adequate. However, the impact of this many simultaneous lightning-like strike disruptions over an extremely large geographic area may exceed those protections. The most significant risk, however, is synergistic because the E2 pulse follows on the heels of the E1. Thus where E1-induced damage has circumvented lightning protection, the E2 impact could pass directly into major system components and damage them.

The late time EMP, or E3, follows E1 and E2 and may last for a minute or more. The E3 pulse is similar in a great many respects to geomagnetic effects induced by solar storms. Solar storms and their impacts on electrical systems with long lines have been thoroughly evaluated and are known to cause serious damage to major electrical system components at much lower levels than the reasonably possible E3 impact. This damage has been incurred in spite of functioning, in-place protective systems. Given the preceding E1 and E2 pulse damage to the protective systems and other system components, damage from E3 to unprotected major system components is virtually assured.

EMP is inimical to the continued functioning of the electrical power system and the reliable behavior of electronics. Each of the three EMP modes of system insult is sufficient by itself to cause disruption and probable functional collapse of large portions of the interconnected electrical power system at EMP threat levels. In every EMP attack, all three assaults (E1, E2, and E3) are delivered in sequence and nearly simultaneously. It is the Commission's assessment that functional collapse of the electrical power system region within the primary area of assault is virtually certain. Furthermore, widespread functional collapse may result even from a small weapon with a significant E1 component. While stopping electrical supply over a broad geographical area nearly instantaneously is damaging, it is the time it takes to restore service that is important, assuming restoration is possible, which itself may be questioned in some instances.

System Collapse Scenarios

NERC was one of several key advisers on the EMP impact assessment discussed above although the conclusions and emphasis are the Commission's alone. NERC also informed the Commission that there is no modeling capability extant, either deterministic or statistical, that can assess with confidence the outcome of simultaneous, combined subsystem failures. Putting together a coherent picture of the projected system collapse scenario must rely on expert judgment.

Large-scale load losses in excess of 10 percent are likely at EMP threat levels. Instantaneous unanticipated loss of load, by itself, can cause system collapse. This is possible at 1 percent loss, and is very likely above 10 percent. At similar percentage levels, loss of generation can also cause system collapse. Both the load loss (normally from a transmission system failure) and generation loss resulting in system collapse have been experienced. At the levels of loss for each, collapse is highly likely if not certain. Systemwide ground-induced currents in the transmission grid can by themselves cause system collapse. They did so in March 1989 in Quebec. At the levels expected in an E3 event, collapse would be much more likely and widespread.

Loss of computer control of substation switchyard equipment could, by itself, lead to system collapse. Manual operation is possible only with adequate communication and the ability of personnel to physically get to the right substations, a problematic question in the event of an EMP attack. Adequate numbers of trained and experienced personnel will be a serious problem even if they could all be contacted and could make themselves available. Thus manual operation would be necessary and might not be timely enough or have sufficient skilled personnel to deal with a broad-scale, instantaneous disruption and dynamic situation. Loss of manual control of switchyard equipment would, in short order, lead to line and transformer faults and trips. Several substations tripping nearly simultaneously would lead itself to system collapse.

Loss of telecommunications would not, by itself, cause immediate system collapse except as needed to address issues caused by the above disruptions. However the lack of telemetered control data would make the system operators effectively blind to what is going on, but personnel at substations, if they can get there and communicate with the system operators, could overcome much of that. Malfunction of protective relays could cause system collapse by contributing to several of the above scenarios through misinformation or by operating incorrectly.

All of these collapse mechanisms acting simultaneously provide the unambiguous conclusion that electrical power system collapse for the NERC region largely impacted by the EMP weapon is inevitable in the event of attack using even a relatively low-yield device of particular characteristics.

Damage Scenarios

The level of damage depends primarily on the functioning of the protective equipment, but it also depends on various aspects of the collapse. In an EMP event, the collapse is virtually instantaneous. The size of the transients on the system may be greater than existing protective systems are capable of handling, even those not damaged by the EMP itself.

Damage to the large transformers and other high-value equipment is directly related to protective relay failure, although it is possible for E1-induced arcs inside transformers to damage transformers irrespective of relay failure. In general, since sequential shutdown is not required, one device per relay is a reasonable rule of thumb. A properly functioning relay has a reasonable chance of protecting the device; an improperly functioning one will probably result in some level of damage in an ensuing system collapse. The level of damage depends on the failure mode. The Commission-sponsored tests were focused on determining the thresholds for damage. EMP threat levels are expected to exceed these thresholds.

Test Results

The EMP Commission conducted both free-field and cable current injection simulation testing. The Commission took the basic stance in its testing program that testing would determine the thresholds at which substantial failure rates (either temporary or permanent) commenced to appear. These, in turn, were used to index attack severities at which the corresponding U.S. infrastructures would be seriously compromised or failed. The Commission's test experience — massively supplemented by that of other U.S. Government operations — was that failure rates typically increased rapidly with peak field amplitude, once a threshold had been attained at which failure or disruption appeared at all. The rationale for such threshold determining testing was that abrupt and synchronized loss of only a few percent of items such as electric power system relays would have grave impacts on the functionality of the system containing such items. This is much like the effect that a few percent of vehicles on a freeway that become disabled would have, producing a serious deleterious impact on the flow of traffic.

A crude rule of thumb was that roughly a factor-of-ten increase in damage effects might be expected when the peak field amplitude was doubled; the exact scaling relation naturally varied from one device type to another and also had very substantial dependence on the frequency content of the pulse, which however, the Commission testing program explored only slightly.

Based on the testing and analysis outlined in this Section, we estimate that a substantial and highly significant fraction of all control and protective systems within the EMP- affected area will experience some type of impact. As the test results were briefed to industry experts at NERC and the Argonne National Laboratory, it became apparent to the Commission that even minor effects noted during the testing could have significant impacts on the processes and equipment being controlled.

Free-Field Testing

EMP free-field simulation testing was conducted using a bounded wave simulator (see **figure 2-4**). The testing was conducted in three phases.

Phase I was dedicated to evaluating the so-called transfer response for the various test systems. This is a measure of the coupling strength of the external perturbation to the test system and provides insight into the expected fraction of the energy in an EMP field that may be deposited into the exposed electronic device. Induced current transients were measured on all the accessible cables of the control systems and measurements were made at the lowest field levels produced by the simulator to minimize the electrical stress on the exposed equipment. Phases II and III of the testing program were focused on obtaining data on fragilities, that is, on identifying the thresholds for induced malfunction or damage response in the tested equipment. A total of eight steps were selected that gradually increased the electrical stress on the control systems. During this testing, all systems were operational with diagnostics and checkouts run at the conclusion of each simulated EMP exposure.

The simulation testing provided an opportunity to observe the interaction of the electromagnetic energy with equipment in an operational mode. Observed effects can be related to the system response in more realistic scenarios through analysis based on coupling differences between the simulated and real-world cases. Since the simulation is not perfect — the pulse length is too long and the test volume too small to capture the longer cable run couplings to be found in a real environment — post test analysis and engineering judgment is required to relate the test results to expected SCADA vulnerabilities in a true EMP event.

There is also no pretense that any test program could possibly do more than selectively sample the wide variety of installed SCADA systems. The choice of representative test systems was guided by findings from previous infrastructure site surveys, by solicited recommendations from industry groups such as the NERC, and by conducting a review of several market surveys.

In the end, four separate control systems were acquired for testing. These systems were representative of those found in power transmission, distribution, power generation, and oil and gas distribution for fueling power plants.

A key observation from this test program is that a wide variety of SCADA, DCS, and PLC malfunctions resulted when exposed to simulated threshold level EMP environments. These ranged from electronic upset of equipment, which might be repaired by either reboot or recycle to physical damage that required the actual replacement of the affected hardware.

The response of the control systems tested varied from system to system as well as from subsystem to subsystem. For example, a unit consisting of multiple input/output ports (or subsystems) connected to a variety of field or communications devices had some ports experience upsets, some experience physical damage, and some experience no effect, all in the same simulation event.

As an example, at relatively low electromagnetic stress levels, a portion of a DCS process controller provided false indications of the process status. An operator interface indicated a switch was on when in actuality it had been turned off, while internal voltage and temperature were reported as out of their normal operating ranges when they were actually normal. These effects were significant because they occurred most frequently on control systems used in SCADA applications, which are geographically dispersed. Correcting these malfunctions typically had to be performed manually at the device location. This approach would greatly complicate the recovery process for geographically distributed systems.

In addition to false readings from the sensors, direct malfunctions of some tested control elements were also noted. Additional control element effects included the failure of pressure transmitters, which included both physical damage and loss of calibration data required to indicate proper readings.

Figure 2-4. EMP Simulator.

Control systems often rely on Ethernet for communications to the man-machine interface as well as communications between controllers in dispersed systems. Communications systems based on Ethernet components similar to those found in PC networking systems suffered substantial degradation and damage effects when illuminated by the simulated albeit low-level EMP pulse. These damage effects are significant since they require the systems to be physically repaired or replaced in order to restore the normal communications capabilities.

Many of the effects noted in the previous paragraphs are attributed to the coupling to the wires and cables interconnecting the systems. The level of this coupling scales roughly with the length of the wire. As a general rule, the larger the transients are in the coupling lines, the more damaging they are to electronics equipment. It is therefore important to consider the transients that might be induced if a more distributed system encounters the same EMP electromagnetic energy. One way to address this concern is to perform cable coupling analysis. This was done as part of the current injection test program in order to relate the susceptibility levels of electronic equipment to the EMP threat.

At the system level, 100 percent of the control systems were affected at times. This is highly relevant to the prospect of system collapse and scope of the problem of restoration. This is more difficult to quantify at the subsystem level due to the sheer number of subsystems associated with each system. Translation to real world conditions must be tempered in cases where the control systems are located in structures that provide electromagnetic shielding of the incident EMP energy, but few of these exist in practice.

Current Injection Testing

Current injection testing was typically done by introducing transient voltage waveforms on a cable leading to the equipment under test. Depending on its load and that of the test generator, current was delivered to the test object. All of the electronics found in the power system is developed using a national (ANSI/IEEE) or international (IEC) standard for a series of electromagnetic compatibility (EMC) waveforms that are representative of the transients observed during normal operation. One waveform that is commonly tested is known as the electrical fast transient (EFT). It has a rise time of 5 ns and a pulse width of 50 ns. By coincidence, this is very similar to the type of waveform coupled to cables by E1.

The objective of this testing was to determine at what level each type of equipment fails to operate normally and also to determine when operator intervention is necessary for the equipment to operate normally. Most of the equipment tested had multiple cable connections, covering different functions (power, signal, communications, etc.). These were all tested.

Given the levels of voltage in which the equipment malfunctioned, a separate effort was performed to compute the coupling of the incident E1 to cables with various lengths and orientations. For the long, exposed cables found in transmission substations, it was found that the induced voltage could exceed 20 kV under many circumstances.

In addition to free-field testing, the Commission sponsored a current injection testing program. The test program was representative in the sense that exemplars of most functional components were tested. Due to expense and time constraints, typically only one or two vendors' equipment was tested, and only one or two samples of a type were tested. The types of equipment tested and results brief are described in the following paragraphs:

Electro-Mechanical Relays

These are the old-fashioned devices that contain no integrated circuits but function using high-power relays. They are still used in about 50 percent of applications, but that share is continuing to decline. As expected, these are immune to EMP upset up to the highest levels tested.

Distribution Line Insulators

Earlier studies have indicated low vulnerability for these simple devices. The Commission-sponsored tests on a variety of 15kV class pin and suspension insulators indicate that there is a higher vulnerability than previously thought New tests performed with the power on found that some insulators were destroyed due to the current following the path of the E1-induced arc. Statistical testing was not performed, so it is not clear what percentage of insulators will behave in this fashion; however, it is clear that power-on testing should be performed in the future to better understand this effect.

Figure 2-5. Test Item: Electronic Relay.

Figure 2-6. Flashover Observed During Injection Pulse Testing.

Electronic Protective Relays

These devices (see **figure 2-5**) are the essential elements preserving high-value transmission equipment from damage during geomagnetic storms and other modes of grid collapse. Fortunately, these test items were the most robust of any of the electronic devices tested. However, test agencies reported that they are subject to upset at higher levels of simulated EMP exposure. We believe that altering the deployment configurations can further ameliorate the residual problems.

Programmable Logic Controllers and Digital Control Systems

These units are most commonly found in industrial settings and in particular are extensively used in power plants. They are subject to upset and damage at moderate levels of EMP assault (see **figure 2-6**). The circuit board pictured is from a typical PLC unit and is exhibiting a damaging short-circuit flashover during EMP Commission-sponsored testing.

General-Purpose Desktop Computers and SCADA Remote and Master Terminal Units

These were the most susceptible to damage or upset of all the test articles. Unlike the other kinds of devices tested, several different models and vintages were examined. The RS-232 ports were found to be particularly susceptible, even at very low levels of EMP stress.

With the exception of the RS-232 connections, all of the electronic devices that were tested performed up to the manufacturer's claimed levels for electromagnetic compatibility. Thus, the international standards to which the manufacturers subscribe are being met. Unfortunately the induced E1 stress is higher than the standards for normal operation.

The net result of this testing provides evidence that the power grid is also vulnerable to collapse due to the E1 component of an EMP assault, primarily through the upset and damage of the soft computer systems that are in common use. This however suggests that operational performance can be considerably enhanced at modest cost by attending to installation and configuration issues.

Historical Insights

To provide insight into the potential impact of EMP-induced electronic system malfunctions, the Commission evaluated previous large service failure events. In these cases, similar (and less severe) system malfunctions have produced consequences in situations that were far too complex to predict using a model or analysis.

Another important observation is that these situations are seldom the result of a single factor but rather a combination of unexpected events, which are easily related to the impact only in hindsight. This is not surprising given the complexity, interdependency, and size of the systems involved. It is important to note that historical examples, while necessary for the insight they provide into the dependence of a functioning modern infrastructure on its automated control systems, do not remotely capture the scale of the expected EMP scenario. In an EMP event, it is not one or a few SCADA systems that are malfunctioning (the typical historical scenario) but very large numbers, hundreds or even thousands over a huge geographic area with a significant fraction of those rendered permanently inoperable until replaced or physically repaired. Critically, the systems that would identify what components are damaged and where they are located are also unavailable in many instances.

Hurricane Katrina, August 2005

Hurricane Katrina, one of the worst U.S. natural disasters ever, caused a widespread, multi-state blackout that lasted for a prolonged period, with catastrophic consequences for the afflicted region. The Katrina blackout was a major factor in the failure of police, emergency and rescue services during the hurricane, which killed 1,464 people. The blackout caused gas stations to cease operating, paralyzing transportation and greatly impeding evacuation efforts. The Katrina blackout, which afflicted the region for weeks and lasted for months in some localities, so severely impeded recovery efforts that even today, 3 years later, New Orleans and its vicinity is still far from being fully recovered.

August 14, 2003, Blackout

The August 14 blackout was precipitated by a single line failure in one control area. It eventually affected nine control areas over a period of several hours, with rapidly spreading cascades of outage over the last 30 minutes. The extent of the blackout was exacerbated by deficiencies in specific practices, equipment, and human decisions. Initial retrospectives have focused on three likely contributory causes:

- Inadequate situational awareness at First Energy Corporation (FEC)
- FEC's failure to adequately manage tree growth in its transmission rights-of-way
- Failure of the interconnected grid's reliability organizations to provide effective diagnostic support.

The inadequate situational awareness and failure to provide effective diagnostic support are closely aligned to the computer and network effects that showed damage and upset during EMP testing. Additionally, new causal features (not common to other blackout incidents) of the August 14 blackout include inadequate interregional visibility over the power system, dysfunction of a control area's SCADA system, and lack of adequate backup capability to that system. Thus, all of the factors involved in the August 14 black- out are expected to be present in control areas impacted by an EMP event, but to a far greater extent. Therefore, an event as large as the ultimate August 14 blackout could be part of an initial EMP impact but multiplied several times over a contiguous geographical and system area. If this effect overlapped the Eastern and Western Interconnections, there is the increased probability that both interconnections could collapse.

Western States Blackout

The 1996 Western States blackout occurred when an electrically loaded transmission line sagged onto a tree and caused a short. This type of event is not uncommon, especially in the heavily treed areas of the Western Interconnect. At about the same time, a second line tripped (opened) due to improper protective relay activation. The tripping of the two transmission lines, coupled with a heavy electrical load on these lines and the thin margins on the transmission system, triggered the widespread outage through cascading failure. An EMP event could be expected to result in the loss of numerous transmission lines at once, not just the two cited in this case.

Geomagnetic Storms

Probably one of the most famous and severe effects from solar storms occurred on March 13, 1989. On this day, several major impacts occurred to the power grids in North America and the United Kingdom. This included the complete blackout of the Hydro-Quebec power system and damage to two 400/275 kV autotransformers in southern England. In addition, at the Salem nuclear power plant in New Jersey, a 1200 MVA, 500 kV transformer was damaged beyond repair when portions of its structure failed due to thermal stress. The failure was caused by stray magnetic flux impinging on the transformer core. Fortunately, a replacement transformer was readily available; otherwise the plant would have been down for a year, which is the normal delivery time for larger power transformers. The two autotransformers in southern England were also damaged from stray flux that produced hot spots, which caused significant gassing from the breakdown of the insulating oil.

The blackout of the Hydro-Quebec system was caused when seven static voltage-amps reactive (VAR) compensators (SVC) tripped and shut down due to increased levels of harmonics on the power lines. The loss of the seven SVCs led to voltage depression and frequency increase on the system, which caused part of the Quebec grid to collapse. Soon afterwards, the rest of the grid collapsed because of the abrupt loss of load and generation. The blackout took less than 90 seconds to occur after the first SVC tripped. About 6 million people were left without power for several hours and, even 9 hours later, there were still 1 million people without power.

Geomagnetic storms represent an approximation to an E3-induced voltage effect. The experience to date is of events that may be orders of magnitude smaller in scope and less severe than that expected from an EMP — although the Commission has also investigated the impact of a 100-year superstorm. The induced geomagnetic superstorm currents in the transmission lines will cause hundreds of high voltage transformers to saturate, creating a severe reactive load in the power system leading to voltage collapse in the affected area and damage to elements of the transmission system. The nature of this threat did not allow for experimental testing of the E3 effect, so this historical record is the best information on the effect.

Distinctions

Past electric power blackouts provide a baseline for assessing the impact of an EMP attack on the power grid as discussed previously. However, there are several important factors that distinguish the EMP collapse scenario from these historical experiences.

- In the historical power system outages, only one or a few critical elements within an entire system have been debilitated. For example, a power generation facility may trip because a surge of current is unexpectedly presented through a fault from a particular load. Yet a substantial portion of the system may well be rendered out of service as the disruption triggers a series of cascading failures, each instigating the next failure (e.g., first a generator trips, then the frequency sags, and a load trips off or a transmission line trips out with its associated loads, which in turn causes the frequency to overrun and another generator trips out, and it continues to oscillate until the interconnected system comes down.) In the case of an EMP attack, elements within many critical facility components are likely to be damaged or disrupted simultaneously over a relatively broad geographic area, thus creating an almost certain cascading collapse of the remaining elements. Similarly, while lightning might strike a single plant, transmission line, or large load causing it to trip out, lightning has not hit multiple locations spread over a very wide area of the system with sufficient intensity and hitting all simultaneously to the extent that would be representative of an EMP attack.
- During historical outages, the telecommunications system and associated control systems have continued to function. This provides the system operators with eyes and ears to know what was damaged, where damage occurred and in some cases the range of damage. While the power system may still come down, it is more possible to take protective measures to minimize damage and impact in order to effectuate rapid restoration. The communications and control systems' functionality are at high risk of disruption and damage themselves during an EMP attack. A minimum communications capability is needed to support immediate responses, to isolate parts for continued operation, and to implement necessary measures to restore the electrical system.
- In the early stages of the EMP attack, even before the disruptions could be sensed and trips could occur that would lead to collapse, some or many of the protective devices will be damaged that have ensured critical system components are safe to allow fast recovery. As a result, some and perhaps much of the electrical system would not be able to protect itself from the effects of multiple simultaneous and cascading failures. Widespread damage to the generation, transmission, and distribution infrastructures and equipment are probable. Rather than simply restoring

power to an intact infrastructure with only a very few damaged components, the recovery task would be to replace an extensively damaged system under very difficult and decaying circumstances and then proceeding to restoration.

- The control systems would be damaged to some extent as opposed to remaining fully operational as in historical outages. The operations and dispatch centers where the vast interconnected system is controlled and managed would probably have damaged and disrupted components, the readings from the system would be fragmented and in many cases false or nonexistent, and communication by whatever means would be difficult to impractical to impossible. Control and knowledge would range from unreliable at best to simply nonexistent. Finding what and where damage has occurred and getting it repaired would be very problematic in any reasonable time frame, even within the control centers themselves, let alone out over the vast network with millions of devices.
- Skilled labor for a massive and diverse repair effort is not currently available if allocated over a large geographic area with great numbers of components and devices to check and repair where necessary. This scope of damage could cover perhaps 70 percent or possibly more of the continental United States as well as a significant part of Canada's population. This is far too large to bring in the limited skilled labor from very distant points outside the affected area in any reasonable time, even if one could coordinate them and knew where to send them, and they had the means to get there. Thus the extensive support from nearby fringe areas used so effectively in historical outages is likely to be unavailable as a practical matter as they themselves would be affected. The blackout resulting from Hurricane Katrina, an event comparable to a small EMP attack, overtaxed the ability of the Nation to quickly restore electric power, a failure that contributed to the slow recovery of the afflicted region.
- Other infrastructures would be similarly impacted simultaneously with the electrical system such as transportation, communication, and even water and food to sustain crews. The ability to find and get spare parts and components or purchase services would be severely hampered by lack of normal financial systems in addition to communication, transportation, and other factors. The Hurricane Katrina blackout caused precisely such problems.
- Fuel supplies for the power generation would be interrupted. First, the SCADA and DCS systems used in delivery of the fuel would be adversely impacted. In addition, much of the fuel supply infrastructure is dependent upon the electrical system. For example, natural gas-fired plants (which make up such a large share of the domestic generation) would be rendered inoperable since their fuel is delivered just in time for use. Coal plants have stockpiles that variously might be adequate for a week to a month. The few remaining oil-fired plants similarly have a limited storage of fuel. Nuclear plants would reasonably be expected to still have fuel but they would have to forego protective regulations to continue to operate. Many renewable fueled resources would still have their fuel supply but EMP effects on controls may still render them inoperable.

It is not possible to precisely predict the time to restore even minimal electrical service due to an EMP eventuality given the number of unknowns and the vast size and complexity of the system with its consequent fragility and resiliency. Expert judgment and rational extrapolation of models and predictive tools suggest that restoration to even a diminished but workable state of electrical service could well take many weeks, with some probability of it taking months and perhaps more than a year at some or many locations; at that point, society as we know it couldn't exist within large regions of the Nation. The larger the affected area and the stronger the field strength from the attack (corollary to extent of damage or disruption), the longer will be the time to recover. Restoration to current standards of electric power cost and reliability would almost certainly take years with severe impact on the economy and all that it entails.

Strategy

The electrical system must be protected against the consequences of an EMP event to the extent reasonably possible. The level of vulnerability and extreme consequence combine to invite an EMP attack. Thus reduction of our vulnerability to attack and the resulting consequences reduces the probability of attack. It is also clear the Cold War type of deterrence through mutual assured destruction is not an effective threat against many of the potential protagonists, particularly those who are not identifiable nation-states. The resulting strategy is to reduce sharply the risk of adverse consequences from an EMP attack on the electrical system as rapidly as possible. The two key elements of the mitigation strategy for the electrical system are protection and restoration.

The initial focus for reducing adverse consequences should be on the restoration of overall electrical system performance to meet critical, if not general, societal needs. The focus should be on the system as a whole and not on individual components of the system. Timely restoration depends on protection, first of high-value assets, protection necessary for the ability to restore service quickly to strategically important loads, and finally protection as required to restore electrical service to all loads. The approach is to utilize a comprehensive, strategic approach to achieve an acceptable risk-weighted protection in terms of performance, schedule, timing, and cost. The effort will include evolution to greater and greater levels of protection in an orderly and cost-effective manner consistent with the anticipated threat level. Where possible, the protection also will enhance normal system reliability and, in so doing, provide great service to society overall.

There is a point in time at which the shortage or exhaustion of critical items like emergency power supply, batteries, standby fuel supplies, replacement parts, and manpower resources which can be coordinated and dispatched, together with the degradation of all other infrastructures and their systemic impact, all lead toward a collapse of restoration capability. Society will transition into a situation where restoration needs increase with time as resources degrade and disappear. This is the most serious of all consequences and thus the ability to restore is paramount.

Protection

It is not practical to try to protect the entire electrical power system or even all high-value components from damage by an EMP event. There are too many components of too many different types, manufactures, ages, and designs. The cost and time would be prohibitive. Widespread collapse of the electrical power system in the area affected by EMP is virtually inevitable after a broad geographic EMP attack, with even a modest number of unprotected components. Since this is a given, the focus of protection is to retain and restore service to critical loads while permitting relatively rapid restoration.

The approach to protection has the following fundamental aspects. These will collectively reduce the recovery and restoration times and minimize the net impact from assault. All of this is feasible in terms of cost and timing if done as part of a comprehensive and reasonable response to the threats, whether the assault is physical, electromagnetic (such as EMP), or cyber.

1. Protect high-value assets through hardening. Hardening, providing for special grounding, and other schemes are required to assure the functional operation of protection equipment for large high-value assets such as transformers, breakers, and generators and to so protect against sequential, subsequent impacts from E2 and E3 creating damage. Protection through hardening critical elements of the natural gas transportation and gas supply systems to key power plants that will be necessary for electrical system recovery is imperative.
2. Assure there are adequate communication assets dedicated or available to the electrical system operators so that damage during system collapse can be minimized; components requiring human intervention to bring them on-line are identified and located; critical manpower can be contacted and dispatched; fuel, spare parts and other commodities critical to the electrical system restoration can be allocated; and provide the ability to match generation to load and bring the system back on line.
3. Protect the use of emergency power supplies and fuel delivery, and importantly, provide for their sustained use as part of the protection of critical loads, which loads must be identified by government but can also be assured by private action. Specifically:
   - Increase the battery and on-site generating capability for key substation and control facilities to extend the critical period allowing recovery. This is relatively low cost and will improve reliability as well as provide substantial protection against all forms of attack.
   - Require key gasoline and diesel service stations and distribution facilities in geographic areas to have at-site generation, fueled off existing tanks, to assure fuel for transportation and other services, including refueling emergency generators in the immediate area.
   - Require key fueling stations for the railroads to have standby generation, similar to that required for service stations and distribution facilities.
   - Require the emergency generator start, operation, and interconnection mechanisms to be EMP hardened or manual. This will also require the ability to isolate these facilities from the main electrical power system during emergency generation operation and such isolation switching must be EMP hardened.
   - Make the interconnection of diesel electric railroad engines and large ships possible and harden such capability, including the continued operation of the units.
   - The Government must determine and specify immediately those strategically important electrical loads critical to the Nation to preserve in such an emergency.
4. Separate the present interconnected systems, particularly the Eastern Interconnection, into several nonsynchronous connected subregions or electrical islands. It is very

important to protect the ability of the system to retain as much in operation as possible through reconfiguration particularly of the Eastern Connected System into a number of nonsynchronous connected regions, so disruptions will not cascade beyond those EMP-disrupted areas. Basically, this means eliminating total NERC region service loss, while at the same time maintaining the present interconnection status with its inherent reliability and commercial elements. This is the most practical and easiest way to allow the system to break into islands of service and greatly enhance restoration timing. This will not protect most within the EMP-insult area, but it should increase the amount of viable fringe areas remaining in operation. This is fiscally efficient and can leverage efforts to improve reliability and enhance security against the broader range of threats, not only EMP. It also can be beneficial to normal system reliability.

5. Install substantially more black start generation units coupled with specific transmission that can be readily isolated to balancing loads. The NERC regions now do surveys of available black start and fuel switchable generation. Requiring all power plants above a certain significant size to have black start or fuel-switching capability (with site-stored fuel) would be a very small added expense that would provide major benefits against all disruptions including nonadversarial ones. Black start generator, operation, and interconnection mechanisms must be EMP hardened or be manual without microelectronic dependence. This also will require the ability to isolate these facilities from the main electrical power system during emergency generation operation and that isolation switching is EMP hardened. In addition, sufficient fuel must be provided, as necessary, to substantially expand the critical period for recovery.
6. Improve, extend, and exercise recovery capabilities. Develop procedures for addressing the impact of such attacks to identify weaknesses, provide training for personnel and develop EMP response training procedures and coordinate all activities and appropriate agencies and industry. While developing response plans, training and coordination are the primary purpose.

Recovery and Restoration

The key to minimizing catastrophic impacts from loss of electrical power is rapid restoration. The protective strategy described is aimed primarily at preserving the system in a recoverable state after the attack, maintaining service to critical loads, and enhancing recovery.

The first step in recovery is identifying the extent and nature of the damage to the system and then implementing a comprehensive plan with trained personnel and a reservoir of spare parts to repair the damage. Damage is defined as anything that requires a trained person to take an action with a component, which can include simply rebooting all the way to replacing major internal elements of the entire component. A priority schedule for repair of generation, transmission, and even distribution is necessary since resources of all types will be precious and in short supply should the EMP impact be broad enough and interdependent infrastructures be adversely impacted (e.g., communication, transportation, financial and life-supporting functions).

Restoration is complicated in the best of circumstances, as experienced in past black-outs. In the instance of EMP attack, the complications are magnified by the unprecedented scope of the damage both in nature and geographical extent, by the lack of information post attack, and by the concurrent and interrelated impact on other infrastructures impeding restoration.

Restoration plans for priority loads are a key focus. Widely scattered or single or small group loads are in most cases impractical to isolate and restore individually given the nature of the electrical system. These are to be served first through the emergency power supply aspects identified in the Protection section. Restoration of special islands can, however, be made practical by the nonsynchronous connected subregions if they are identified by the Government as necessary very far in advance of any assault. Otherwise, the system's resources and available personnel will need to act expeditiously to get as many islands of balanced load and generation back into operation. This will begin by system operators identifying those easiest to repair (normally the least damaged) and restore them first. As these stabilize, the system recovery will flow outward as, increment by increment, the system is repaired and brought back in service. It is much more feasible and practical to restore by adding incrementally to an operating island rather than black starting the recovery for an island.

Balancing an isolated portion of generation and load first, and then integrating each new increment is a reasonably difficult and time-consuming process in the best of circumstances. In an EMP attack with multiple damaged components, related infrastructure failures, and difficulty in communications, restoring the system could take a very long time unless preparatory action is taken.

Generating plants have several advantages over the widely spread transmission network as it relates to protection and restoration from an EMP event. The plant is one complete unit with a single DCS control network. It is manned in most cases so operators and maintenance personnel are immediately available and on site. The operating environment electronically requires a level of protection that may provide at least a minimal protection against EMP. Nevertheless, it is important to harden critical controls sufficiently to enable manual operation at a minimum. Providing for at-site spares to include the probably needed replacements for control of operation and safety would be straightforward and not expensive to accomplish, thus assisting rapid restoration of capability.

As controls and other critical components of the electrical transmission and generation system suffer damage, so do similar components on the production, processing, and delivery systems providing fuel to the electric generators. Restoration of the electrical power system is not feasible on a wide scale without a parallel restoration of these fuel processing and delivery systems.

Hydropower, wind, geothermal, and solar power each has a naturally reoccurring fuel supply that is unaffected by EMP. However, the controls of these plants themselves are subject to damage by EMP at present. In addition, only hydropower and geothermal have controllable fuel (i.e. they can operate when needed versus wind and solar that operate when nature provides the fuel just-in-time). As a practical matter, only hydropower is of sufficient size and controllability in some regions to be a highly effective resource for restoration, such as the Pacific Northwest, the Ohio/Tennessee valley, and northern California. Beyond the renewable resources, coal and wood waste plants typically have significant stockpiles of fuel so the delay in rail and other delivery systems for a couple of weeks and in some instances up to a month is not an issue for fuel. Beyond that, rail and truck fuel will be needed and delivery times are often relatively slow, so the delivery process must start well before the fuel at the generator runs out.

Operating nuclear plants do not have a fuel problem per se, but they are prohibited by regulation from operating in an environment where multiple reliable power supply sources are not available for safe shutdown, which would not be available in this circumstance. However, it is physically feasible and safe for nuclear plants to operate in such a circumstance since they all have emergency generation at site. It would simply have to be fueled sufficiently to be in operation when the nuclear plant is operating without external electrical supply sources. Nuclear power backup would need to be significantly expanded. Natural gas-fired power plants are very important in restoration because of their inherent flexibility and often their relatively small size, yet they have no on-site fuel storage and are totally dependent upon the natural gas supply and gas transportation system which are just in time for this purpose. Therefore, the natural gas fuel delivery system must be brought back on-line before these power plants can feasibly operate. It is operated largely with gas turbines of its own along the major pipelines. The key will be to have the protection, safety, and controls be hardened against EMP.

Recovery from transmission system damage and power plant damage will be impeded primarily by the manufacture and delivery of long lead-time components. Delivery time for a single, large transformer today is typically one to two years and some very large special transformers, critical to the system, are even longer. There are roughly 2,000 transformers in use in the transmission system today at 345 kV and above with many more at lesser voltages that are only slightly less critical. No transformers above 100 kV are produced in the United States any longer. The current U.S. replacement rate for the 345 kV and higher voltage units is 10 per year; worldwide production capacity of these units is less than 100 per year. Spare transformers are available in some areas and systems, but because of the unique requirements of each transformer, there are no standard spares. The spares also are owned by individual utilities and not generally available to others due to the risk over the long lead time if they are being used. Transformers that will cover several options are very expensive and are both large and hard to move. NERC keeps a record of all spare transformers.

Recovery will be limited by the rate of testing and repair of SCADA, DCS, and PLC and protective relay systems. With a large, contiguous area affected, the availability of outside assistance, skilled manpower, and spares may well be negligible in light of the scope of the problem. Information from power industry representatives enables us to place some limits on how long the testing and repair might take. Determining the source of a bad electrical signal or tiny component that is not working can take a long time. On the low side, on-site relay technicians typically take three weeks for initial shakedown of a new substation. Simply replacing whole units is much faster, but here too, inserting new electronic devices and ensuring the whole system works properly is still time consuming. It must be noted that the substations are typically not manned so skilled technicians must be located, dispatched, and reach the site where they are needed. Many of these locations are not close to the technicians. It is not possible to readily estimate the time it will take in the event of an EMP attack since the aftermath of an EMP attack would not be routine and a certain level of risk would likely be accepted to accelerate return to service. It seems reasonable, then, to estimate an entire substation control system recovery time to be at least several days, if not weeks. This assumes that the trained personnel can reach the damaged locations and will be supported with water, food, communication, spare parts, and the needed electronic diagnostic equipment.

Unlike generation, recovery of the transmission system will require off-site communications because coordination between remote locations is necessary. Communications assets used for this purpose now include dedicated microwave systems and, increasingly, cell phones and satellite systems. If faced with a prolonged outage of the telecommunications infrastructures, repairs to dedicated communication systems or establishment of new ad-hoc communications will be necessary. This might take one or more weeks and would set a lower limit on recovery time, but it would be unlikely to affect the duration of a months-long outage.

Restoration to electrical service of a widely damaged power system is complex. Beginning with a total blackout, it requires adequate communication to match and coordinate a generating plant to a load with an interconnected transmission that normally can be isolated via switching at several substations, so it is not affected by other loads or generation. The simultaneous loss of communication and power system controls and the resulting lack of knowledge about the location of the damage all greatly complicate restoration. There are also a diminishing number of operators who can execute the processes necessary for restoration without the aid of computers and system controls.

Without communication, both voice and data links, it is nearly impossible to ascertain the nature and location of damage to be repaired, to dispatch manpower and parts, and to match generation to load. Transportation limitations further impede movement of material and people. Disruption of the financial system will make acquisition of services and parts difficult. In summary, actions are needed to assure that difficult and complex recovery operations can take place and be effective in an extraordinarily problematic post-attack environment.

The recovery times for various elements of the electrical system are estimated in the following paragraphs. These should be regarded as very rough best estimates for average cases derived from the considered judgment of several experts. These estimates are gross averages, and the situation would vary greatly from one facility to another as the situation, number of disrupted and damaged elements and the extent of preassault preparedness and training vary. In addition, the contingencies and backlogs strongly depend on the extent of such damage elsewhere and are essentially unknown. For example, fuel delivery capability is a key element. Each of the system elements — generation (including fuel delivery), transmission, distribution, and often load — must be repaired and in working order sufficient for manual control at a minimum (each element with skilled personnel all in communication with each other). Thus, the following should be occurring in parallel as much as possible, but in some instances testing of one element requires a working capability of another. The availability of spare parts and trained manpower coupled with knowledge of what to repair and where it is are critical to recovery timing. The recovery times provided below are predicated upon the assumption that the other infrastructures are operating normally. The recovery times would increase sharply with the absence of other operating infrastructures, which is likely in the EMP situation. These estimates are based upon present conditions, not what is possible if the Commission recommendations are followed.

Power Plants

- Replace damaged furnace, boiler, turbine, or generator: one year plus production back-log plus transportation backlog. It is uncertain if and to what extent damage to these elements will occur if the protection schemes are disrupted or damaged.
- Repair some equipment if spares on site exist, but repair time depends on the type of plant and personnel available at the plant at the time of the assault: two days to two weeks plus service backlog at the site or to move trained personnel from plant to plant.
- Repair and test damaged SCADA, DCS, and computer control system: three months.
- Return repaired or undamaged plant to operation, provided the major components under the first bullet are not damaged: (1) nuclear: three days provided there is an independent power feed with enough fuel, which should be on site in such an emergency, (2) coal: two days plus black start or independent power feed, (3) natural gas: two hours to two days depending on fuel supply and black start, (4) hydro: immediate to one day, (5) geothermal: one to two days, (6) wind: immediate to one day unless each turbine requires inspection and then one or two turbines a day.
- All of the above are also contingent on the availability of fuel. Our recommendations for on-site reserves: coal: 10-30 days; natural gas: depends on whether the pipeline is operating; nuclear: 5 days to several weeks; hydro: depends on reservoir capacity available for continued use.

Transmission and Related Substations

- Replace irreparably damaged large transformers: One to two years plus production backlog plus transportation plus transportation backlog (these are very large and require special equipment to transport that may not be available in this situation).
- Repair damaged large transformer: one month plus service backlog.
- Repair manual control system: one month if adequate personnel are available.
- Establish ad hoc communications: one day to two weeks.
- Repair and test damaged protective systems: three months.
- Repair and return of substations to service are also contingent on the local availability of power. All substations have batteries for uninterrupted power, nominally enough for eight hours. Very few (about 5 percent) have on-site emergency generators. Many utilities rented emergency generators in advance of the Y2K transition. Almost all are now gone. Once the local power is gone, other emergency power often must be brought to the station for operation.
- Assuming DC terminals are manned: one week to one month depending upon damage.

Distribution and Related Substations

- Replace insulators that have flash-over damage: two to five days, unless very widespread and then weeks.

- Replace service transformers: two to five days unless very widespread and then weeks.
- Repair time depends on the number of spares, available crews that perform the repairs, and equipment.
- Note that the load on the end of the distribution may have some disruption that needs repair as well.

Starting an electrical power system from a fully down and black system requires one of the following two approaches. (1) At the margin of the outage, an operating electrical system is running at proper frequency with balanced load to generation, and this system can be interconnected to the fringe of the black portion of the system. The newly inter-connected portion, the portion being restored, must be able to sequentially (in increments or simultaneously) bring on load and generation to keep the now larger portion of the system in sufficient frequency balance so the entire system, new and old, does not collapse. Then another increment is integrated into the operating system and so on. As the portion that is operating in balance becomes larger and more flexible, the increments that are able to be added become larger as well, since the operating system can absorb more and maintain stability. This is how historical outage areas are predominantly restarted. (2) There is generation that can be black started. This means starting a generator without an external power source, such as hydroelectric or diesel generation. To do this, the generation has to be synchronized on line, and a load has to be matched to the generation as it comes on line. That requires that a transmission link between the generation and the load be put in service. Yet the transmission link also must be segregated from the rest of the system, or the load hanging on it would be too large. Both approaches require that the increment of the system being re-energized to be fully functional (repaired) and communication established between the generation and load, including any substation or switch- yard between the two. Importantly, it requires skilled personnel to execute the restoration manually.

The generation must have sufficient fuel to accommodate the load being met in balance. Water behind a hydroelectric facility may be limited and certainly the diesel fuel is likely to be limited. Thus the startup must be done carefully because failures could render the black start inoperable as it runs out of fuel or depletes the battery. Normally, in this type of situation, the diesel or small hydro is used primarily to start up a larger generator of a size that can carry the necessary load increment. This larger generator must be fueled, which can be a complication as discussed elsewhere in this Section.

Under deregulation, the disconnection (in the business sense) of transmission from generation that has been occurring in the U.S. electrical power business creates a problem for black start recovery. There are risks involved in returning a plant to operation and costs for the needed repairs. Questions about who will pay whom and who will follow whose direction is not easy to answer, even with everybody wanting to cooperate. Under the historic utility monopolies, the generation and the transmission assets had a common owner, so these matters were handled within a single organization. Now, coordination with independent power producers is nearly unenforceable other than through heavy government emergency powers noting that power producers and owners want indemnification before assuming risk. Therefore some degree of command authority is required for coordination, assessment, and acceptance of risk of damage and financial settlement of losses.

The time to integrate sufficient portions of a black region of the system using the fringe approach is reasonably short if the outage area is small in relation to the operating area as has been seen in past outage conditions. In the case of EMP, where the outage area is likely to be much larger than the fringe area or there is no fringe area, restoration of even parts will be measured in weeks to months. If communication is difficult to nonexistent, restoration can take much longer.

Mitigation of Adverse Consequences

By protecting key system components, structuring the network to maximize fringe service, through the nonsynchronous interconnections, expanding the black start and system emergency power support, creating comprehensive recovery plans for the most critical power needs, and providing adequate training of personnel, the risk of catastrophic impact to the Nation can be significantly reduced. The mitigation plan must be jointly developed by the Federal Government and the electric power industry, instilled into systems operations, and practiced to maintain a ready capability to respond. It must also be fully coordinated with the interdependent infrastructures, owners, and producers.

The continuing need to improve and expand the electric power system as a normal course of business provides an opportunity to judiciously improve both security and reliability in an economically acceptable manner — provided that technically well-informed decisions are made with accepted priorities. There are a wide variety of potential threats besides EMP that must be addressed, which can have serious to potentially catastrophic impacts on the electrical system. Common solutions must be found that resolve these multiple vulnerabilities as much as possible. For example, in the course of its work, the Commission analyzed the impact of a 100-year solar storm (similar to E3 from EMP) and discovered a very high consequence vulnerability of the power grid. Steps taken to mitigate the E3 threat also would simultaneously mitigate this threat from the natural environment. Most of the precautions identified to protect and restore the system from EMP will also apply to cyber and physical attacks. The Commission notes that the solutions must not seriously penalize our existing and excellent system but should enhance its performance wherever possible.

The time for action is now. Threat capabilities are growing and infrastructure reinvestment is increasingly needed which creates an opportunity for the investment to serve more than one purpose. Government must take responsibility for improvements in security. As a general matter, improvements in system security are a Government responsibility, but it may also enhance reliability if done in certain ways. For example, providing spare parts, more black start capability, greater emergency back-up, nonsynchronous interconnections, and more training all will do so. Yet, EMP hardening components will not increase reliability or enhance operation. Conversely improving reliability does not necessarily improve security, but it may if done properly. For example, adding more electronic controls will not enhance EMP security, but electronic spare parts and more skilled technicians will help improve security and reliability. Finding the right balance between the utility or independent power producer's service and fiscal responsibility with the Government's security obligation as soon as possible is essential, and that balance must be periodically (almost continuously) reexamined as technology and system architecture changes.

Recommendations

EMP attack on the electrical power system is an extraordinarily serious problem but one that can be reduced below the level of a catastrophic national consequence through focused effort coordinated between industry and government. Industry is responsible for assuring system reliability, efficiency, and cost effectiveness as a matter of meeting required service levels to be paid for by its customers. Government is responsible for protecting the society and its infrastructure, including the electric power system. Only government can deal with barriers to attack — interdiction before consequence. Only government can set the standards necessary to provide the appropriate level of protection against catastrophic damage from EMP for the civilian sector. Government must validate related enhancements to systems, fund security-only related elements, and assist in funding others.

It must be noted, however, that the areas where reliability and security interact represent the vast majority of cases. The power system is a complex amalgamation of many individual entities (public, regulated investor-owned, and private), regulatory structures, equipment designs, types and ages (with some parts well over one hundred years old and others brand new). Therefore, the structure and approach to modifications must not only recognize the sharply increased threat from EMP and other forms of attack, but improvements must be accomplished within existing structures. For example, industry investment to increase transmission capacity will improve both reliability and system security during the period when transmission system operating margins are increased.

The Commission concluded that mitigation for a majority of the adverse impact to the electrical system from EMP is reasonable to undertake in terms of time and resources. The specific recommendations that follow have been reviewed with numerous entities with responsibility in this area. The review has been in conceptual terms, with many of the initiatives coming from these parties, but the recommendations are the Commission's responsibility alone. The activities related to mitigation of adverse impacts on fuel supply to electric generation are more fully discussed in a separate Section of this report.

Responsibility

As a result of the formation of Department of Homeland Security (DHS) with its statutory charter for civilian matters, coupled with the nature of EMP derived from adversary activity, the Federal Government, acting through the Secretary of Homeland Security, has the responsibility and authority to assure the continuation of civilian U.S. society as it may be threatened through an EMP assault and other types of broad scale seriously damaging assaults on the electric power infrastructure and related systems.

It is vital that DHS, as early as practicable, make clear its authority and responsibility to respond to an EMP attack and delineate the responsibilities and functioning interfaces with all other governmental institutions with individual jurisdictions over the broad and diverse electric power system. This is necessary for private industry and individuals to act to carry out the necessary protections assigned to them and to sort out liability and funding responsibility. DHS particularly needs to interact with FERC, NERC, state regulatory bodies, other governmental institutions at all levels, and industry in defining liability and funding relative to private and government facilities, such as independent power plants, to contribute their capability in a time of national need, yet not interfere with market creation and operation to the maximum extent practical.

DHS, in carrying out its mission, must establish the methods and systems that allow it to know, on a continuous basis, the state of the infrastructure, its topology, and key elements. Testing standards and measurable improvement metrics should be defined as early as possible and kept up to date.

The NERC and the Electric Power Research Institute (EPRI) are readily situated to provide much of what is needed to support DHS in carrying out its responsibilities. The Edison Electric Institute, the American Public Power Association, and the North American Rural Electric Cooperative Association are also important components for coordinating activity. Independent power producers and other industry groups normally participate in these groups or have groups of their own. The manufacturers of generation, transmission, and distribution components are another key element of the industry that should be involved. Working closely with industry and these institutions, DHS should provide for the necessary capability to control the system in order to minimize self- destruction in the event of an EMP attack and to recover as rapidly and effectively as possible.

Multiple Benefit

Most of the recommended initiatives and actions serve multiple purposes and thus are not only to mitigate or protect against an EMP attack and other assaults on the electric power system. The protection of the system and rapid restoration of the system from an EMP attack also are effective against attack from a number of physical threats that directly threaten to destroy or damage key components of the electrical system. Large- scale natural disasters, such as Hurricane Katrina, also are in large part mitigated by these same initiatives. Many of the initiatives enhance reliability, efficiency, and quality of the electrical supply, which is a direct benefit to the electrical consumer and the U.S. economy.

To the greatest extent feasible, solutions for EMP should be designed to be useful solutions to the broad range of security and reliability challenges. For example, black start resources are essential for many threats, purposeful or not, to the power grid. Integrating cyber security and EMP hardness into control systems simultaneously as these systems are routinely upgraded will be much more effective and less costly than doing two separate jobs.

Recommended Initiatives

The following initiatives must be implemented and verified by DHS and DOE, utilizing industry and other governmental institutions to assure the most cost effective outcome occurs and that it does so more rapidly than otherwise possible. In many instances, these initiatives are extensions or expansions of existing procedures and systems such as those of NERC.

- *Understand system and network level vulnerabilities, including cascading effects*—To better understand EMP-related system response and recovery issues, conduct in-depth research and development on system vulnerabilities. The objective is to identify cost effective and necessary modifications and additions in order to further achieve the overall system performance. Specifically there should be government-sponsored research and development of components and processes to identify and develop new consequential and cost effective approaches and activities.
- *Evaluate and implement quick fixes*—Identify what may presently be available commercially to provide cost effective patches and snap-on modifications to quickly

provide significant protection and limit damage to high-value generation and transmission assets as well as emergency generation and black start capability. These include installation or modification of equipment as well as changes in operating practices. This is both fast and low cost.

- *Develop national and regional restoration plans*—The plans must prioritize the rapid restoration of power with an emphasis on restoring critical loads that are identified by the Government. The plans must be combined with the requirements for providing and maintaining emergency power service by these loads. The plans must address outages with wide geographic effect, multiple component failure, poor communication capability, and failure of islanding schemes within the affected area. Government and industry responsibilities must be assigned and clearly delineated. Indemnification arrangements must be put into place to allow industry to implement the Government's priorities as well as deal with potential environmental and electrical hazards to ensure rapid recovery. Planning must address not only the usual contingency for return to normal operating condition, but also restoration to a reduced capability for minimum necessary service. Service priorities under duress may be different from priorities under normal conditions. The planning basis for reduced capability should be the minimum necessary connectivity, generation assumptions based on reduced fuel availability scenarios, and reduced load, with the goal of universal service at limited power. National Guard and other relevant resources and capabilities must be incorporated.
- *Assure availability of replacement equipment*—On hand or readily available spare parts to repair or replace damaged electronic and larger power system components must be available in sufficient quantities and in locations to allow for rapid correction and restoration commensurate with a post-EMP attack and its impacts on related infrastructures such as communication and transportation. NERC already has a spare component database for such large items as transformers and breakers that is expanding to include delivery capability, but now must be revised to accommodate an EMP attack environment. Where additional spare components need to be acquired or delivery made possible to critical locations, DHS must work with NERC and industry to identify the need and provide the spares or delivery capability; such as the critical material and strategic petroleum reserves and similar strategic reserves. The key will be to decide where to draw the line between reserves for reliability and those for security. It also will be necessary to keep the equipment current. In addition, strategic manufacturing and repair facilities themselves might be provided with emergency generation to minimize stockpiles. This would also be of benefit to industry as well as enhance security. Research is underway and should be further pursued, into the production of multiple use emergency replacement transformers, breakers, controls, and other critical equipment. Such devices would trade efficiency and device service life for modularity, transportability and affordability. They would not be planned for normal use. Movement, stockpiles and protection of stockpiles must be integrated with National Guard and other relevant capabilities.
- *Assure availability of critical communications channels*—Assure that throughout the system there are local and system-wide backup EMP survivable communication systems adequate for command and control of operations and restoration of the electrical system. The most critical communications channels are the ones that enable

recovery, not normal operations. Planning must presume that, for the near term at least, computer-based control systems will not be capable of supporting post-EMP operations. The most critical communication assets are thus the in-house ones that enable manual operation and system diagnostics. Dispatch communication is next in importance. Communications to coordinate black start are also vital. NERC should review and upgrade operating procedures and information exchanges between and among existing control centers, key substations, and generating plants to recognize and deal as effectively as possle with EMP, building upon the systems, procedures, and databases currently in place. Local emergency and 9-1-1 communications centers, the National Guard and other relevant communication systems, and redundant capabilities should be incorporated where possible.

- *Expand and extend emergency power supplies*—Add to the number of stand-alone back-up and emergency power supplies such as diesels and long-life batteries. This addition is vital and a least-cost protection of critical service. The loss of emergency power before restoration of the external power supply is likely to occur in present circumstances and is highly probable to be devastating. Presently such emergency power is useable only for relatively short periods due mostly to at-site stored fuel limitations, which have become increasingly limited. The length of time recommended for each location and load will be determined by DHS and industry where the emergency supply is private, such as with hospitals, financial institutions, and telecommunication stations. The specific recommendations are:
  - Increase the battery and on-site generating capability for key substation and control facilities to extend the critical period allowing recovery. This action is relatively low cost and will improve reliability as well as provide substantial protection against all forms of attack.
  - Require key gasoline and diesel fuel service stations and liquid fuel distribution facilities in geographic areas to have at-site generation, fueled from existing at-site storage to assure fuel for transportation and other services, including refueling emergency generators in the immediate area.
  - Require that key fueling stations for the railroads have standby generation much as the previously mentioned service stations and distribution facilities.
  - Require the emergency generator start, operation, and interconnection mechanisms to be EMP hardened or manual. This action will also require the ability to isolate these facilities from the main electrical power system during emergency generation operation and require that such isolation switching be EMP hardened.
  - Where within safety parameters extend the emergency generation life through greater fuel storage or supply sources (with their own emergency power supplies). Fuel supplies for more critical facilities must be extended to at least a week or longer, where possible. This action will probably entail careful use or development of relatively near location (but not contiguous) fuel stockpiles with their own emergency generation.
  - Regularly test and verify the emergency operations. If the Government were to enforce current regulations, many of the public facilities with standby generation would be routinely tested and failures could be avoided.

  - Provide for the local integration of railroad mobile diesel electric units with switching and controls hardened against EMP. The same should be provided for large ships at major ports.
- *Extend black start capability*—Systemwide black start capabilities must be assured and exercised to allow for smaller and better islanding and faster restoration. The installation of substantially more black start generations units and dual feed capable units (e.g., natural gas-fired units that can operate on #2 oil stored on site) coupled with specific transmission that can be readily isolated to balance loads for restoration is necessary. Sufficient fuel must be provided to substantially expand the critical period for recovery such as with multiple start attempts. The NERC regions now do surveys of available black start and fuel switchable generation. Requiring all power plants above a certain significant size to have black start or at-site fuel switching capability (with at-site stored fuel) would be a very small added expense, and would provide major benefits against all disruptions including nonadversarial, so it is both an industry and security benefit. The start, operation, and control systems for such capability have to be EMP hardened or manual, recognizing that most large power plants have personnel on site.
- *Prioritize and protect critical nodes*— Government entities, such as DHS and DOE, must identify promptly those specific loads that are critical to either remain in service or to be restored as a priority with target restoration to be within a matter of hours following an EMP attack. These may well include loads necessary to assure the continuation of all forms of emergency response care and recovery. These must include what is necessary to avoid collapse of, or allow for the rapid recovery of financial systems, key telecommunication systems, the Government's command and control in the civilian sector, and those elements that allow for rapid and effective recovery of the electric power system in a more general sense. These loads must be prioritized so that the most critical can be protected and designed for rapid restoration in the near term and then add more next-level priority loads as resources permit. The above recommendations for extended and adequate emergency power supply are the most direct and cost efficient approach. The shift to nonsynchronous, interconnected islands is the secondary application, but it will take longer and is more expensive. Providing such islands of small-to-modest size to support large loads can best assure no loss of power supply or far more rapid restoration.
- *Expand and assure intelligent islanding capability*—Direct the electrical system institutions and entities to expand the capability of the system to break into islands of matching load and generation, enhancing what now exists to minimize the impact and provide for more rapid and widespread recovery. The establishment of nonsynchronous connections between subregions, perhaps beginning with NERC already identified subregions, should be required. This can readily be accomplished today with approaches such as DC back-to-back converter installations that facilitate power transfers but maintain a barrier. This mode of operation between regions is often referred to as maintaining frequency independence. Reconfiguration of the Eastern Connected System into a number of such nonsynchronous connected regions could eliminate large service interruptions, while still maintaining the present interconnection status. It may be a priority to first establish smaller islands of frequency independence to better assure power supply to government-identified

critical loads that are nominally too large for most emergency power supplies, such as large financial centers, and tele-communication hubs. Incidental to any studies could be new ideas for conversion of HVAC transmission lines to HVDC operation for greater transmission capacity as a further and corollary benefit. Also new ideas are being discussed, such as, where the converter transformers can be eliminated, resulting in a substantial cost reduction. Asynchronous regional connections is a common term used to identify this broad area technically. The protective and control systems necessary to implement this capability will have to be hardened. It will not be a retrofit but simply a part of the initial design and procedures, so the cost for EMP protection is small. Note that the DC or other interface making the nonsynchronous connection possible is not sized for the entire electrical capacity within the respective island but is sufficient only for reliability and commercial transactions, which normally is far less. Sizing this interface is a special effort that needs to be established primarily by NERC and FERC but with Federal coordination. Breaking the larger electrical power system into subsystem islands of matching load and generation will enhance what now exists to minimize the impact, decrease likelihood of broad systemwide collapse, and provide for more rapid and widespread recovery. It is just as useful for normal reliability against random disturbances or natural disasters in reducing size and time for blackouts. Thus it is critical for protection and restoration coming from any type of attack, not just EMP. Ensuring this islanding capability in the event of EMP is critical, although it requires a longer-term system design and implementation.

- *Assure protection of high-value generation assets*—Enhance the survivability of generating plants at the point of system collapse due to the very broad and simultaneous nature of an EMP attack. NERC, EPRI, equipment and control system providers, and utilities need to aggressively evaluate and verify what is vulnerable to EMP and commensurate consequences. Generating plants can be severely damaged from large electrical faults or incursions in the absence of protective devices. They can also be occasionally damaged in the event of sudden load loss if protective shutdown systems fail. Control systems used in generation facilities are inherently less robust than their counterparts in transmission and thus are more susceptible to EMP disruption. They are highly computer controlled which further exacerbates their risk to EMP. Yet at the same time, they have trained personnel on site who with proper training, procedures, and spare parts, can greatly assist in restoration. System-level protection assurance is more complex due to the need for multiple systems to function in proper sequence. Lead times on generation components are even longer than for major transmission components. Existing coal plants make up nearly half the Nation's generation, but they generally have the most robust control systems with many remaining electro-mechanical controls still in operation. Natural gas-fired combustion turbines and associated steam secondary systems represent the newest significant contribution to the generation. These are mostly all modern electronic- and computer-based control and protective systems and are considered very vulnerable to EMP. Their fuel systems are not on site and will also be interrupted due to EMP. Nuclear plants have many redundant and fail-safe systems, but they too are very electronically controlled. The key difference with nuclear power plants is the extensive manual control capability and training, making them less vulnerable than the others. Hydroelectric is the next substantial generation element and is the most robust, although its older mechanical and electromechanical controls are being replaced at a rapid rate. Black start generation is normally quite secure but start and frequency controls will need to be protected from EMP. The highest priority generation assets are those needed for black start, but all are critical for restoration of any meaningful service.

- *Assure protection of high-value transmission assets*—Ability to withstand EMP must be assured at the system level. Priority for protection is on the highest voltage, and on the highest power units serving the longest lines; these require the most time to replace and are the most vulnerable in the absence of normal protections due to E1 and provide the major flow and delivery of power. Provisions must be made for the protection of large high-value assets such as transformers and breakers against the loss of protection and sequential subsequent impacts from E2 and E3 creating damage. E3 ground-induced current impacts are important from an industry standpoint since they can occur beyond E3 due to the risk of large, 100-year geomagnetic solar storms. For E3 this could include adding either permanent or switchable resistance to ground in the neutral of large transformers. This protection would then be available upon notice of the onset of a solar storm or sufficient threat of EMP attack. Thus it provides a simple expedient that does not compromise performance under normal operation. Due to the interconnected nature of the grid and to the need for that connectivity to enable recovery, the likelihood of a blackout lasting years over large portions of the affected region is substantial with damage to these high-value components. The islanding of the system through nonsynchronous connections may help reduce the E2 and E3 impacts by shortening the long line coupling in some instances.
- *Assure sufficient numbers of adequately trained recovery personnel*—Expand levels of manpower and training as they are otherwise limited to only that needed for efficient normal power operation that is highly and increasingly computer aided. Industry and government must work together to enhance recovery capability.
- *Simulate, train, exercise, and test the recovery plan*— Develop two or three centers for the purpose of simulating EMP and other major system threatening attacks. Develop procedures for addressing the impact of such attacks to identify weaknesses, provide training for personnel and develop EMP response training procedures and coordination of all activities and appropriate agencies and industry. While developing response plans, training and coordination are the primary purpose, identifying vulnerabilities through "red team" exercises is also important for identifying, prioritizing, and rectifying weaknesses. The centers would each focus on one of the three main integrated electrical networks — Eastern Grid, Western Grid, and Texas. These centers may be able to effectively utilize facilities such as the TVA bunker and the BPA control center in order to conserve resources and achieve rapid results. DOE facilities and other no longer utilized facilities should also be examined. Develop simulators to train and develop procedures similar to the airline industry. Exercising black start will require indemnification of power providers.
- *Develop and deploy system test standards and equipment*—Test and evaluate the multitude of system components to ensure that system vulnerability to EMP is identified and mitigation and protection efforts are effective. Device-level standards

and test equipment exist for normal power line disturbances (EMC standards), but protection at the system level is the more important goal. System-level improvements such as isolators, line protection, and grounding improvements will be the most practical and least expensive in most cases rather than replacement of individual component devices.

- *Establish installation standards*—More robust installation standards must be identified and implemented as appropriate — such as short shielded cables, circumferential grounding, arrestors on leads, surge protectors, and similar activities. These should include more robust system standards — such as proximity to protected device, no commercial off-the-shelf (COTS) computers in mission critical roles and similar matters. In some instances, these will qualify as add-ons and replacements during the early period initiatives. The Government should complete the testing and evaluation work that the Commission initiated to set hardening standards for electric power protective systems. Government should provide fiscal assistance to industry in implementing the needed hardening solutions.

Cost and Funding of Selected Initiatives

It must be noted that the very wide variety of components; installation techniques; local system designs; age of components, subsystems, and controls located within buildings or exposed; and so forth all drastically affect the type and expense for implementing the recommended initiatives. Internal DHS and other governmental costs are assumed to be absorbed. A significant portion of the labor to affect the modifications is already in place. Often the modification will be part of a program for repair, replacement and modernization that is continuing regardless of the EMP mitigation program. The addition of non-synchronous connection capability once defined is a contract function coupled with at- site staffing and control system interfaces. All of this effort factors into the cost estimates and results in fairly wide ranges in most instances. Only the costs for some of the larger or more system-specific initiatives are estimated here (in 2007 dollars).

- There are several thousand major transformers and other high-value components on the transmission grid. Protective relays and sensors for these components are more than that number but less than twice. A continual program of replacement and upgrade with EMP-hardened components will substantially reduce the cost attributable uniquely to EMP. Labor for installation is already a part of the industry work force. The estimated cost for add-on and EMP-hardened replacement units and EMP protection schemes is in the range of $250 million to $500 million.
- Approximately 5,000 generating plants of significance will need some form of added protection against EMP, particularly for their control systems. In some instances the fix is quite inexpensive and in others it will require major replacements. The estimated cost is in the range of $100 million to $250 million.
- The addition of nonsynchronous interfaces to create subregion islands is not known with reasonable certainty, but it might be in the order of $100 million to $150 million per island. The pace of creating islands and their priority will be established by DHS in consultation with NERC and FERC. Moving to at least six or more fairly rapidly is

a fair assumption. There will be annual operating costs of around $5 million per island.

- The simulation and training centers are assumed at three — one for each interconnect — for a cost in the range of $100 million to $250 million plus annual operating costs of around $25 million per year.
- Protection of controls for emergency power supplies should not be too expensive since hard-wired manual start and run capability should be in place for many, which is adequate. Furthermore, the test, adjust, and verification will be carried out by the entity that owns the emergency power supply as part of normal operating procedures. Retro-fit of protective devices such as filters might be accomplished at a cost of less than $30,000 per generator for newer generators with vulnerable electronic controls. Hardening the connection to the rest of the facility power system requires a protected internal distribution system from the backup generator.
- Switchable ground resistors for high-value transformers are estimated to cost in the range of $75 million to $150 million.
- The addition of new black start generation with system integration and protected controls is estimated to cost around $12 million per installation. Probably no more than 150 such installations will need to be added throughout the United States and Canadian provinces. Adding dual fuel capability to natural gas-fired generation is done for the economic purpose of the owner, yet it has the same value as the addition of black start generation. The addition of fuel storage for the existing black start units is relatively small, about $1 million each.
- The addition of emergency generation at the multitude of sites including fuel and transportation sites is probably around $2 million to $5 million each.
- The cost for monitoring, on a continuous basis, the state of the electric infrastructure, its topology, and key elements plus for assessing the actual EMP vulnerability, validation of mitigation and protection, maintenance, and surveillance data for the system at large cannot be estimated since it falls under many existing government-funded activities, but in any event, it is not considered significant.
- Research and development activities are a level-of-effort funding that needs to be decided by DHS. Redirection of existing funding is also likely to occur.
- Funding for the initiatives above is to be divided between industry and government. Government is responsible for those activities that relate directly and uniquely to the purpose of assuring continuation of the necessary functioning of U.S. society in the face of an EMP attack or other broadly targeted physical or information systems attack. Industry is responsible for all other activities including reliability, efficiency and commercial interests. Industry is also the best source for advice on cost effective implementation of the initiatives.

SECTION 3. TELECOMMUNICATIONS

Introduction

Telecommunications provides the connectivity that links the elements of our society together. It is a vital capability that plays an integral role in the normal day-to-day routine of the civilian, business, and government sectors of society. It is a critical enabler for the functioning of our national financial infrastructure, as transactions representing trillions of dollars flow daily via telecommunications. It enables agencies of local, state, and federal government to discharge their duties. People can communicate on the go, almost anytime and virtually anywhere because of telecommunications, as exemplified by more than 100 million cellular subscribers in the United States (U.S.). Telecommunications provides a vital pathway between emergency response personnel in crisis situations. It has transformed, via the Internet and advances in technology, the way business and society in general operate. Downloading music and video content using the Internet instead of in-store purchases, using cell phones to interactively gather travel directions instead of using paper maps, and using remote sensors and video streams to send security information over a communications network to a central site for appropriate dispatch instead of using on-site security guards are examples of these changes.

Telecommunications can be thought of as:

- The mix of equipment used to initiate and receive voice, data, and video messages (e.g., cell phones and personal computers).
- The associated media (e.g., fiber optics and copper) and equipment (e.g., multiplexers) that transport those messages.
- The equipment that routes the messages between destinations (e.g., Internet Protocol [IP]-based routers).
- The basic and enhanced services offered by communications carriers such as AT&T, Verizon Wireless, and Comcast.
- The supporting monitoring and management systems that identify, mitigate, and repair problems that can impact performance of services.
- The supporting administrative systems for functions such as billing.

This Section discusses civilian telecommunications. Among the main trends to consider in evaluating the impact of EMP on these telecommunications networks in the next 15 years are:

- The dramatic growth in the number of wireless networks and in the use of wireless services.
- Improvements in the technology and reliability associated with optical networks lever- aging heavy fiber deployment (fiber is generally viewed positively in terms of EMP survivability).
- Shrinking work forces used in managing networks and an associated increase in dependence on automation and software "diagnostic smarts" to support maintenance, problem isolation and recovery, and other performance impacting functions.

- An architectural evolution toward a converged network in which voice, data, and video traffic are carried over the same network.

When fully implemented, this evolution to a converged network will represent a major change-out of the equipment that existed in the 1990s, and that still exists, in the U.S. telecommunications network. Thus, it represents an opportunity for EMP hardening considerations to be included as the transition occurs.

Telecommunications service providers have proclaimed that carrying voice, data, and video together over converged networks is an underpinning of their strategic directions. Service providers point to the fact that traffic residing on embedded, older technology will be transitioned to this new converged network within financial and regulatory constraints.[5] While this converged network evolution has begun, it is expected to continue for an additional decade or more.

The reason for a lengthy transition can be better understood by reviewing some historical factors related to the U.S. telecommunications network. Several factors led to traffic being carried by separate networks, including differences in the characteristics of voice, data, and video traffic; the relative dominance in the amount of voice traffic over data and video; and the technological state of the carrier network equipment.

With respect to traffic characteristics:

- Voice communications generally are characterized by real-time interactions with typical durations of a few minutes.
- Data communications tend to occur in bursts and may consume large amounts of bandwidth during these bursts. Data communications users often access networks for long holding times that may range into hours.
- Video traffic typically is characterized by high-bandwidth, long-duration, one-way transmission such as distributing cable TV content to viewers with lower-bandwidth traffic sent from the subscriber to the service provider, for example, to signal the selection of a specific on-demand program.

With respect to traffic mixes:

- As the 1990s progressed, the growth of data traffic exploded, fueled in large part by Internet usage. Data communications growth is continuing at a rapid pace, while growth in voice has remained relatively flat. Some estimates have data traffic already exceeding voice traffic beginning around the year 2000.
- The growth in data communications traffic has made it more fiscally attractive to find technological solutions that avoid the expense of maintaining separate voice and data networks.

With respect to technology evolution:

- Voice communications over the past decade have been handled primarily by carrier equipment called digital circuit switches. The switches are engineered based on statistical usage of the network that assumes not all of the individual users, traditionally in the thousands, served by those individual switches will try to gain access simultaneously. These switches, of which several thousand are deployed, were not designed to effectively handle the characteristics of data and video traffic.

- Router technology has evolved rapidly. Advances in protocols that support assigning quality of service (QoS) requirements for different traffic mixes and greater processing speed and capacity have provided solutions for handling voice, data, and video using a common set of equipment.
- Service providers implement new technologies slowly. This is prudent given the complexity of the networks in question and a desire to prove-in technologies and fine-tune network management procedures prior to wide-scale deployment.

Telecommunications Support during Emergencies

There is a recognition at the highest levels of government and industry that telecommunications plays a critical role, not only in the normal day-to-day operations of society, but also in reconstituting societal functions and mitigating human, financial, and physical infrastructure losses during man-made and natural disasters. This has led to government and industry partnering to codify processes, organizational structures, and services to address these disasters. Among these codifications are the National Communications System (NCS) and a set of services known as National Security and Emergency Preparedness (NS/EP) services.

The NCS was established by Executive Order 12472, *Assignment of National Security and Emergency Preparedness Telecommunications Functions*.[6] These functions include administering the National Coordinating Center for Telecommunications (NCC) to facilitate the initiation, coordination, restoration, and reconstitution of NS/EP telecommunications services or facilities under all crises and emergencies; developing and ensuring the implementation of plans and programs that support the viability of telecommunications infrastructure hardness, redundancy, mobility, connectivity, and security; and serving as the focal point for joint industry-government and interagency NS/EP telecommunications planning and partnerships.

With respect to the NS/EP telecommunications services, a set of evolving capabilities exist for:

- Prioritizing telephone calls through the wireline and wireless networks during time intervals when call volumes are excessive and facilities may be degraded.
- Giving priority to restoring emergency and essential services that may be damaged or degraded.
- Rapidly getting new telecommunications connections into operation.
- Keeping carriers communicating with government and one another on an on-going basis during crises events.

NS/EP-related definitions are noted below.

NS/EP DEFINITIONS

NS/EP Telecommunications Services—Telecommunications services that are used to maintain a state of readiness or to respond to and manage any event or crisis (local, national, or international) that causes or could cause injury or harm to the population, damage to or loss of property, or degrades or loss of property, or degrades or threatens the NS/EP posture of the United States. *(Telecommunications Service Priority [TSP] System for National Security Emergency Preparedness: Service User Manual, NCS Manual 3-1-1, Appendix A, July 9, 1990)*

NS/EP Requirements—Features that maintain a state of readiness or respond to and manage an event or crisis (local, national, or international), which causes or could cause injury or harm to the population, damage to or loss of property, or degrade or threaten the NS/EP posture of the United States. (Federal Standard 1037C)

Emergency NS/EP and Essential NS/EP—Emergency NS/EP telecommunication services are those new services that are ''so critical as to be required to be provisioned at the earliest possible time without regard to the costs of obtaining them.'' An example of Emergency NS/EP service is federal government activity in response to a Presidential declared disaster or emergency. Telecommunications services are designated as essential where a disruption of ''a few minutes to one day" could seriously affect the continued operations that support the NS/EP function. (Federal Register/Vol. 67, No. 236, December 9, 2002/Notices)

These NCS and NS/EP services are capabilities that would be drawn upon in an EMP event, and they will evolve as the U.S. telecommunications network evolves. This commitment to evolution has been reinforced, for example, by testimony from Frank Libutti (Undersecretary, Information Analysis and Infrastructure Protection, Department of Homeland Security) before the United States Senate Committee on Appropriations in 2004:

> The NCS is continuing a diverse set of mature and evolving programs designed to ensure priority use of telecommunications services by NS/EP users during times of national crisis. The more mature services—including the Government Emergency Telecommunications Service (GETS) and the Telecommunications Service Priority (TSP)—were instrumental in the response to the September 11 attacks. FY 2005 funding enhances these programs and supports added development of the Wireless Priority Service (WPS) program and upgrades to the Special Routing Arrangement Service (SRAS). Specifically, priority service programs include: (1) GETS, which offers nationwide priority voice and low-speed data service during an emergency or crisis situation; (2) WPS, which provides a nationwide priority cellular service to key NS/EP users, including individuals from Federal, state and local governments and the private sector; (3) TSP, which provides the administrative and operational framework for priority provisioning and restoration of critical NS/EP telecommunications services. In place since the mid-1980s, more than 50,000 circuits are protected today under TSP, including circuits associated with critical infrastructures such as electric power, telecommunications, and financial services.; (4) SRAS, which is a variant of GETS to support the Continuity of Government (COG) program including the reengineering of SRAS in the AT&T network and development of SRAS capabilities in the MCI and Sprint networks, and; (5) the Alerting and Coordination Network (ACN) which is an NCS program that provides dedicated communications between selected critical government and telecommunications industry operations centers.[7]

EMP Impact on Telecommunications

To aid in understanding the impact of EMP on telecommunications, **figure 3-1** provides a simplified diagram of a telecommunications network.

Service subscribers communicate through a local node. For example, a cellular subscriber communicates through a cell tower controlled by a cellular base station. If communication is with a party located on another local node, the communications traffic may be routed through the backbone to the distant local node for delivery to the other party. The backbone connects to thousands of local nodes and in doing so serves a transport and routing function to move voice, data, or video traffic between or among the communicators. It consists of a mix of equipment that provides high-speed connectivity between the local nodes. In an actual network if there is sufficient traffic between two local nodes, they may be directly connected by transmission media such as fiber links. **Figure 3-1** shows some network equipment, such as a digital switch and a network router. The control network collects information statistics from the equipment in the local nodes and backbone that help manage the network's performance. The backbone has been the main focus of industry in deploying components of the converged network to date, and it is the furthest along with respect to the converged network vision.

A set of first-order assumptions drove the analytical assessment of EMP impacts on telecommunications:

- In a crisis, voice services will be viewed as critical, with the percent of call attempts completed as a key metric.
- The backbone, as depicted in **figure 3-1**, is where the greatest influx of new equipment has been deployed. This is newer-vintage, expensive, high-end routing and transport equipment connected by fiber optics. An assumption is that the equipment will be highly survivable up to high E1 EMP levels and perhaps will experience only transient effects at those levels, but this needs to be verified with further testing.
- The local nodes will be replaced with equipment supporting the converged network vision, but this change-out will continue beyond the time frame examined in this Commission study. Commission-sponsored testing provided insights into the performance of the new equipment that is being incorporated into the converged network. Among the current local node equipment are digital circuit switches and other equipment that have been tested and analyzed as part of a prior assessment of E1 EMP on telecommunications conducted in the early 1990s.[8] In this study, circuit switch manufacturers noted they would be incorporating equipment changes to address a majority of the items shown to be susceptible to E1 EMP in the products tested, and the Commission assessments assume this to be the case.

Keeping these factors in mind, the Commission focused its analytical efforts on customer premises equipment (CPE), the subsequent impact on demand levels at the local nodes and local node equipment, and the subsequent ability to complete calls assuming a robust backbone.

On the demand side, call origination electronic assets have the potential for EMP disruption or damage. A key issue is whether EMP will impact the operation of telephones, cell phones, and computer systems (like those shown in **figure 3-1**) and, as such, reduce the demand placed on assets in the local and backbone elements that move information between information senders and receivers.

The major elements of the civilian telecommunication network are electronic systems with circuit boards, integrated circuit chips, and cable connections such as routers that switch and transport information between users of the network (e.g., transport phone calls). Like the equipment that generates demand on the network, these electronics have an inherent vulnerability to EMP threats. The majority of these critical switching and transport assets that are part of the local and backbone nodes in **figure 3-1** are housed in Central Offices (COs). Typically COs are windowless concrete buildings. Sometimes equipment used to provide service to end users is housed in Controlled Environmental Vaults (CEV). These are smaller structures that provide environmental control similar to that of a CO. Wireless base stations supporting cellular communications are housed in structures similar to CEVs. Finally, some equipment such as that used to provide high-speed Internet service may be installed in small cabinets and enclosures without environmental controls.

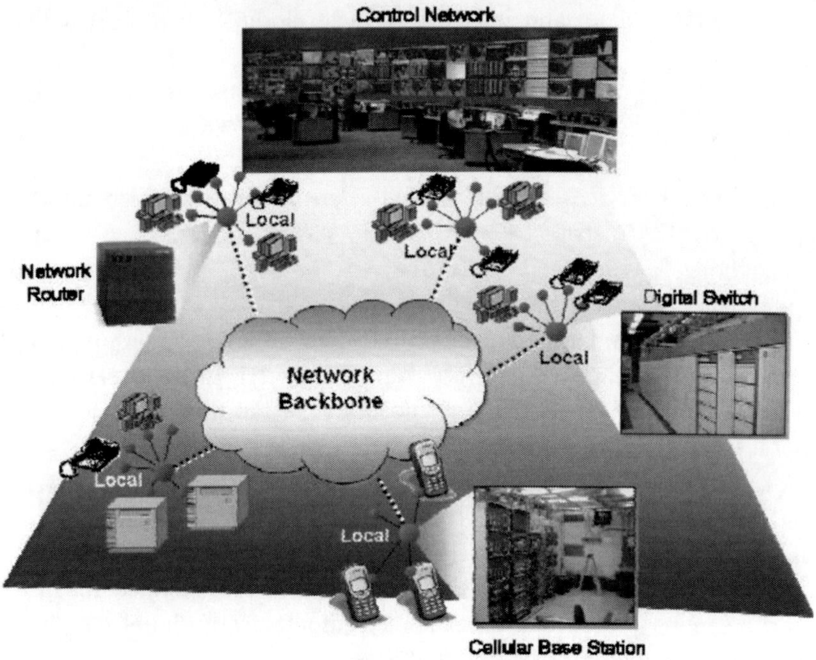

Figure 3-1. Generic Telecommunications Network Architecture.

Regardless of the installation location, telecommunications equipment and the facilities that contain them follow strict rules and requirements to protect against natural or unintentional electromagnetic disturbances, such as lightning, electromagnetic interference, electrostatic discharge, and power influences on telecom cables. Typical protection techniques include grounding, bonding, shielding, and the use of surge protective devices. However, an EMP attack exhibits unique characteristics, such as rapid rise-time transients, and the existing protection measures were not specifically intended for or tested against EMP.

Given these network characteristics, some factors that contribute to mitigating EMP effects on telecommunications are:

- Industrywide groups that systematically share best practices and lessons learned to improve network reliability, such as the Network Reliability and Interoperability Council (NRIC).
- Availability of NS/EP telecommunications capabilities.
- Volume, geographic diversity, and redundant deployment of telecommunications equipment assets, coupled with wireline, wireless, satellite, and radio as alternative means for communications.
- Deployment of fiber-optic technology within telecommunications carrier networks.
- Use of standard bonding and grounding practices for telecommunications equipment deployed in carrier networks.
- Historical performance of terrestrial carrier networks in electromagnetic events such as lightning and geomagnetic storms.

The Commission sponsored testing and analytical efforts that led to the conclusion that an EMP attack would disrupt or damage a functionally significant fraction of the electronic circuits in the Nation's civilian telecommunications systems in the geographic region exposed to EMP. Cellular networks are seen as being less robust to EMP than landline networks due to a combination of the higher susceptibility of cellular network equipment to damage and more limited backup power capacity at cell sites than at counterpart landline network equipment sites.

The analysis suggested that damage to telephones, cell phones, and other communications devices would not be sufficient to curtail higher than normal call volumes on the civilian telecommunications network after exposure to either low or high E1 EMP levels. As such, the remaining operational network would be subjected to high levels of call attempts for some period of time after the attack, leading to degraded telecommunications services. Key government and nongovernment personnel will need priority access to use public network resources to coordinate and support local, regional, and national recovery efforts. This will be especially problematic during the interval of severe network congestion. Services such as GETS will be crucially important during these periods of high call demand.

The Commission's expectation is that the impact of a low E1 EMP level exposure would be dominated by the inability to handle the spike in call traffic on landline networks, because the direct impacts on equipment are expected to be largely transient and short term in nature (minutes to hours) with minimal manual restoration. For cellular networks, the impact will be greater (minutes to days) due to the expected levels of manual recovery, more limited backup power at cell sites, and the large number of cellular base stations that serve as key controllers of communications between cell towers and cell phones. The results of limited testing on cellular base stations indicate EMP vulnerabilities that require further examination.

As noted in the electric power section of the Commission report, the loss of portions of the power grid is likely, even for a relatively low-level EMP attack. The longer-term performance of the public telecommunications network and associated NS/EP services will depend, therefore, on the use of backup power capabilities and the rapidity with which primary power can be restored. To offset a loss of electric power, telecommunication sites now use a mix of batteries, mobile generators, and fixed-location generators. Typically, these have 4 to 72 hours of backup power available on-site and thus will depend on either the resumption of electrical utility power or fuel deliveries to function for longer periods of time. A short-term electric power grid outage (less than a few days) would not cause a significant loss of telecom services due to the existence of power backup systems and best practices supporting these critical systems.

In the case of high amplitude E1 EMP level exposures, spikes in call traffic, coupled with a mix of transient impacts and damage requiring manual network equipment restoration, will result in degraded landline and cellular communications on the order of days to weeks. As in the case of low E1 levels, longer-term impacts from power outages could extend the period and severity of the degradation.

General results from the Commission's EMP analysis received concurrence from the NCS as noted below.

Senate Testimony (March 2005)

In March 2005 testimony before a U.S. Senate subcommittee (Terrorism and the EMP Threat to Homeland Security, Subcommittee on Terrorism, Technology, and Homeland Security, March 8, 2005), the Acting Director of the NCS noted that "Just last year, the NCS also actively participated in the congressionally-chartered *Commission to Assess the Threat from High Altitude Electromagnetic Pulse* (the 2004 EMP Commission) that examined and evaluated the state of the EMP threat at present and looking 15 years into the foreseeable future. The Commission's Report, delivered last July, concludes that EMP presents a less significant direct threat to telecommunications than it does to the National Power grid but would nevertheless disrupt or damage a functionally significant fraction of the electronic circuits in the Nation's telecommunications systems in the region exposed to EMP (which could include most of the United States). The NCS concurs with this assessment."

Analysis Approach

To estimate the impact of an EMP attack on the civilian telecommunications network, the following major tasks were performed:

- Reviewed lessons learned with respect to telecommunications critical dependencies and susceptibilities from past studies of events, such as major disasters and Year 2000 (Y2K).
- Visited telecommunications facilities to get "ground truth" insights into possible areas of EMP susceptibility and for data such as equipment layouts to support illustrative testing of telecommunications equipment.
- Reviewed past test data and performed illustrative testing of wireline and cellular communications devices such as cell phones and network equipment such as network routers to determine EMP susceptibilities.
- Developed models of telephone network restoration processes and network call processing associated with alternative EMP scenarios using subject matter expert judgment, illustrative test data, and augmentation of existing models to estimate degradation levels for networks. Network statistics such as call completion levels

used to estimate degradation were generated for users, assuming they were not using NS/EP services such as GETS.

Analysis Approach—Lessons Learned

From interviews and reviews of lessons learned from past outage events, the following issues were identified that helped shape Commission recommendations and provided input for the testing and modeling activities:

- Y2K contingency planning and past outage events, such as the Hurricane Katrina blackout, point to the need for a functioning voice communications network in an emergency situation to support restoration efforts for multiple critical infrastructures. For example, with respect to managing the power grid, reference material associated with Y2K preparations noted, "The principal strategy is to operate using a manual transfer of a minimum set of critical information … electric systems must provide sufficient redundancy to assure voice communications over a geographic area that addresses its critical facilities and interfaces to neighboring systems and regional centers."[9]
- Conditions that would lead to multi-day unavailability of power remain a principal concern of telecommunications providers. Extended power outages will exacerbate attempts to repair damage and lead to fuel shortages that end up taking network capacity off-line. This concern was reinforced by Hurricane Katrina and by the August 2003 Northeast Power Outage. The latter was a key topic of the August 27, 2003, NRIC meeting.
- A high level of call attempts on both wireline and wireless networks will follow an EMP attack, thereby reducing the effectiveness of voice communications for some time period. At least four times the normal call traffic will likely be experienced by these networks. In previous disasters, these high levels generally lasted for 4 to 8 hours and remained slightly elevated for the first 12 to 24 hours after the event. The spike in call volumes results in callers experiencing problems in successful call completion. Additionally, callers may experience conditions such as delayed dial tones or "all circuits busy" announcements. As an example, the high blocking levels experienced by callers on cellular networks on September 11, 2001, in Washington, D.C., and New York City are shown in **figure 3-2** as call attempts rose to levels as high as 12 times normal.[10]
- As previously noted, concerns over the ability of key personnel to get calls through the public telecommunications network in a disaster was one of the catalysts for services development that occurred under the leadership of the NCS. GETS and WPS are services intended for use in emergency conditions to improve the probability of key personnel completing calls even when wireline and wireless network are under extremely heavy call loads. These services will be leveraged during an EMP event, but their benefits for subscribers are mitigated when local equipment requires manual recovery to be functional. Based on test results, this manual recovery requirement for cellular base stations is of particular concern.

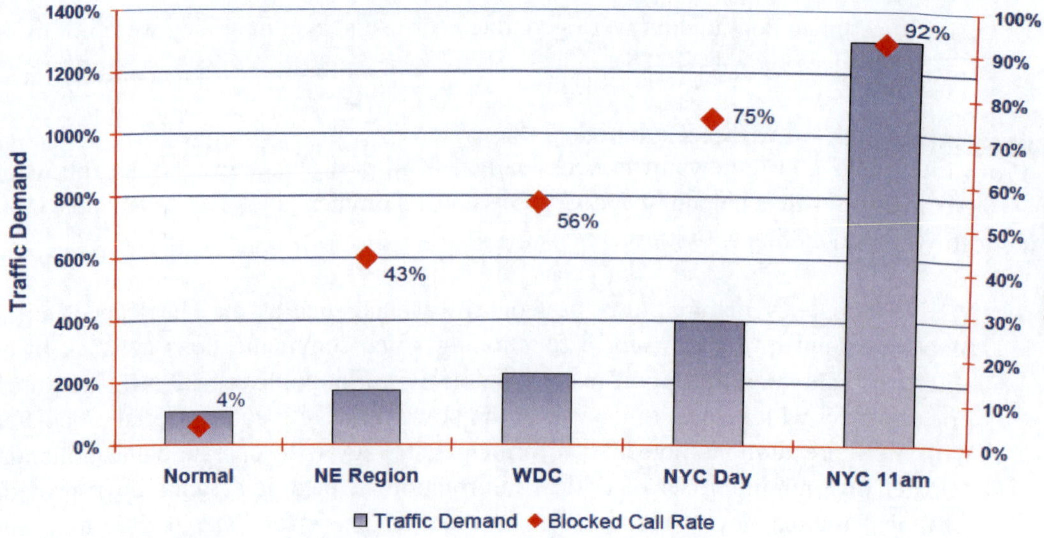

Figure 3-2. September 11, 2001, Blocked Call Rate—Cellular Networks.

- Maintenance and control functions will be critical to restoration and recovery efforts, as they are used by telecommunications carriers to alleviate the overload conditions and identify areas of damage within the network to hasten recovery efforts. For the general populace without access to NS/EP services, if massive call attempts tie up network resources there would be minimal circuits available to dial out and potentially reduced capability to reach 9-1-1 services. To help alleviate this, personnel in a Network Management Center (see **figure 3-3**) could issue a command to the carrier network for "call gapping" through a few quick keystrokes on a personal computer. Through this command, some percentage of calls would be stopped at the originating switch and thus free up resources that would be needed for dialing out. Testing conducted as part of the previously referenced NCS-sponsored assessment indicated that some physical damage to circuit switch components linking to these network management facilities would occur, even at very low transient levels. This damage would reduce the ability of recovery efforts to bring systems up to full capacity and affect the ability to remotely implement procedures to address EMP-induced network problems.

Figure 3-3. Example Network Management Facility.

Analysis Approach—Collecting Ground Truth

Prior to conducting testing on equipment, visits were made to carrier facilities to verify some of the assumptions regarding equipment layouts that were used in the test configurations. Sites containing wireline network switching and transport equipment, cellular network switching and transport equipment, and Network Management Center equipment were visited. Features such as cable lengths and bonding and grounding practices and issues such as policies for stockpiling spares were explored during these visits. In addition, discussions were held with personnel involved in telecommunications equipment installation activities, technical requirements development for electromagnetic effects protection, and network monitoring and control activities to vet assumptions made in the equipment testing and modeling activities. **Figure 3-4** shows cellular network base station equipment photographed during one of the visits.

Figure 3-5 is a photo of router equipment used to collect performance information from carrier equipment and transmit it to a Network Management Center, such as the one in **figure 3-3**.

Based on the collected data, a process for network restoration was developed considering the wide mix of assets that could be affected in an EMP event. The restoration process was reviewed with experts who had been involved in large restoration efforts, including personnel charged with developing software systems to expedite network recovery. These reviews helped augment the restoration process model. This process was used in developing recovery timelines generated in the modeling and simulation activity.

Figure 3-4. Cellular Base Station Equipment.

Figure 3-5. Routers Collecting Network Management Data.

Analysis Approach—Testing for EMP Effects on Telecommunications Networks

Using the lessons learned and ground truth data described previously, a test plan was developed that focused most heavily on the effects of EMP on voice communications and the associated maintenance and control networks that would support recovery and restoration efforts. Consistent with this, testing activities focused on communications devices and switching and routing equipment expected to play a critical role in supporting future voice communications and on computing equipment supporting the collection of data used for network traffic management. E1 was considered as the primary source of EMP effects on carrier equipment under the assumption that long transport lines within carrier telecommunications networks have moved to fiber instead of copper. We also recognized the growing use of fiber within close proximity to home and business establishments. In accordance with these assumptions, the communications carrier network equipment testing focused on assets that would be considered part of the local nodes in **figure 3-1**.

Prior test data on digital switches, routers, computers, and related equipment were reviewed. For example, during the 1980s and early 1990s, the NCS sponsored testing on major telecommunications switches and transport equipment. The test effort conducted on behalf of the EMP Commission was designed to complement the data available in previously discussed NCS technical reports and other data sources. The test data provided information on the behavior of particular pieces of equipment and was subsequently used to model the impact of an EMP attack on the telecommunications network infrastructure and the recovery process. In addition to network equipment, CPE such as basic telephones and cell phones were tested, as the level of traffic on the public telecommunications networks would be affected by the CPE's EMP survivability. **Table 3-1** lists the telecommunications assets tested at multiple government and commercial facilities, including a rationale for why they were selected. A mixture of continuous wave immersion (CWI), pulse current injection (PCI), and free field illumination tests was used. **Figure 3-6** depicts testing that was conducted at a cellular base station at Idaho National Laboratory (INL). Free-field illumination testing was conducted on equipment covering each of the areas in **table 3-1** (except for cellular network carrier switching equipment [see **figure 3-6**]). The equipment tested included a softswitch, cordless phones, cellular phones, computing servers, Ethernet switches, and routers.

During the testing, in cases where impacts were observed, some were transient in nature, for example, auto-rebooting of softswitch equipment, while some testing resulted in permanent equipment damage and required manual recovery via replacement of components (for example Ethernet card replacement) to address performance degradation.

Table 3-1. Telecommunications Equipment Tested

Items	Importance
Corded Phones, Cordless Phones, Cell Phones	Key devices used for voice communications. The level of demand placed on the public telecommunications network will be impacted by the equipments' operational state.
Computing Servers, Secure Access Devices	These computers house software supporting key management and control functions (Network Fault and Traffic Management) critical to network recovery efforts. Since these systems may have to be accessed remotely in an emergency, secure access devices that generate passwords are used to gain access to them.
Routers, Ethernet Switches	Critical equipment supporting the routing of network control and status information between network elements and the facilities and computer systems responsible for their management.
Softswitches, Gateways	Key equipment being integrated into public networks to support the trans- mission of voice, data, and video over IP-based technology. This equipment is replacing the digital circuit switches that are part of the local nodes shown in **figure 3-1**.
Mobile Switching Centers, Base Stations, Base Station Controllers	Major operational components of cellular networks that are used to transmit cellular calls.
Cable Modem Termination System (CMTS), Cable Modems	Cable companies are moving aggressively into telecom, and cable modems are heavily used by customers to access the cable network for communications. The CMTS converts the data signals from cable modems to an Internet Protocol. Trends point to the increased use of routers, Ethernet switches, softswitches, and gateways to route communications traffic.

Figure 3-6. Cellular Network Testing at INL.

Figure 3-7. Testing at NOTES Facility.

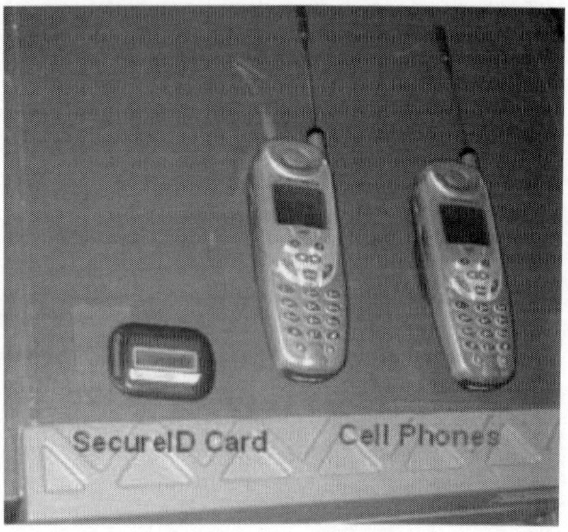

Figure 3-8. Secure Access Card and Cell Phones.

Figure 3-8. shows examples of some of the smaller items tested at the NOTES facility.

Analysis Approach—Modeling and Simulation of EMP Effects

To develop a view of the system effects that would be caused by an EMP attack, a systematic approach was used in the modeling and simulation effort. The analysis leveraged the Commission-sponsored testing just described, as well as prior equipment testing results. Initially, a telecommunications network performance modeling approach for generating call completion levels given degradation assumptions in wireline and wireless networks was developed for the continental United States. The major assumption in this modeling was that the key area of degradation would be local nodes in the carrier networks (for both wireline and cellular networks). As shown in **figure 3-1**, local nodes are equipment such as the digital circuit switches and cellular base station equipment that provide callers with entry into these wireline and cellular telecommunications networks. Impacts on local nodes could inhibit local calls, as well as prohibit connections to the backbone network that provides for more geographically dispersed communications. Positive trends in the direction of EMP survivability for backbone communications are due to increased routing diversity coupled with heavy fiber deployment, suggesting that a local focus is reasonable in terms of first-order effects.

Following this logic, the modeling steps included:

1. Generate a case study using weapons detonation scenarios that produce electromagnetic field levels modeled over selected geographic regions of the United States and model the impact on network performance (e.g., call completion levels) given the degree of network upset expected to be caused initially by the EMP event. We included transient or self-correcting effects and effects that require human action to correct. The model incorporated past test results from NCS studies and new testing of the equipment listed in **table 3-1**, using assumptions about the types and configurations of equipment that would be deployed in affected areas. The starting point for equipment types was industry databases identifying equipment deployed in telecommunications networks. This was augmented with subject matter expert discussions.
2. Apply generic methods and procedures incorporated in the network restoration process noted earlier to generate recovery times for network equipment. Inputs include engineering assumptions on equipment damage levels, availability of repair personnel, availability of network management and control functions, availability of electric power, and other factors.
3. Use the recovery times to model placing equipment back in service and iteratively estimate network performance levels over time using the network performance model.

The following are illustrative results generated from among scenarios of interest identified by the Commission. **Figures 3-9** through **3-11** show originating call completion levels in the eastern United States for the average combined wireline and wireless calls after an EMP event. Time period categories in these figures include immediately following, 4 hours after, and 48 hours after the EMP event. The results displayed incorporate longer restoration times for cellular equipment, driven in part by levels of manual recovery. **Figure 3-12** shows the recovery curve during the 10-day period following the attack. This is the estimated time period to regain pre-event performance, absent other infrastructure interdependency impacts such as long-term power outages. The shaded circles indicate EMP field-level isocontours generated by the weapon. For example, in **figure 3-9**, the geographic area most negatively impacted has estimated call completion levels of roughly only 4 percent, while the area outside the range of the direct effects has a 73 percent call completion level estimate.

The reason for the 73 percent level is that callers outside the directly affected areas are unable to make calls into the affected areas due to equipment disruptions in those areas, coupled with network congestion and high call-retry levels.

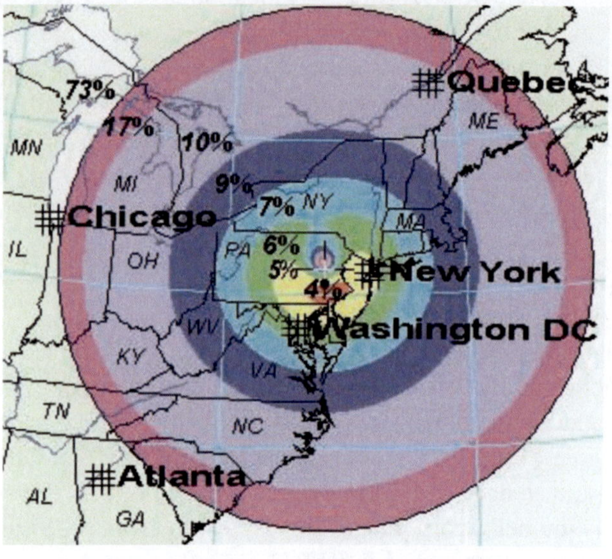

Figure 3-9. Percentage of Calls Completed Immediately After EMP Event.

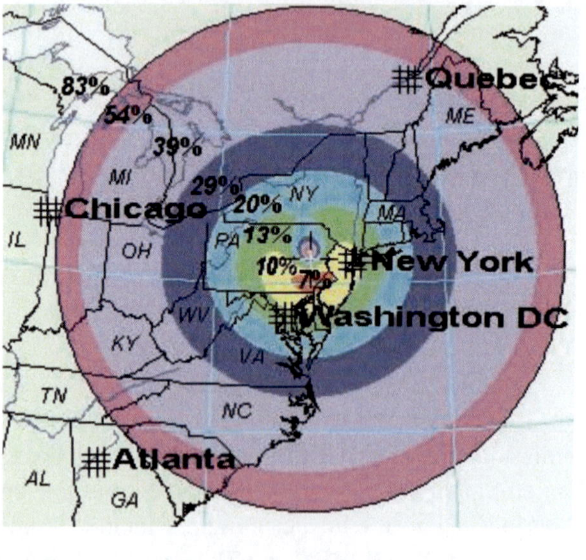

Figure 3-10. Percentage of Calls Completed 4 Hours After EMP Event.

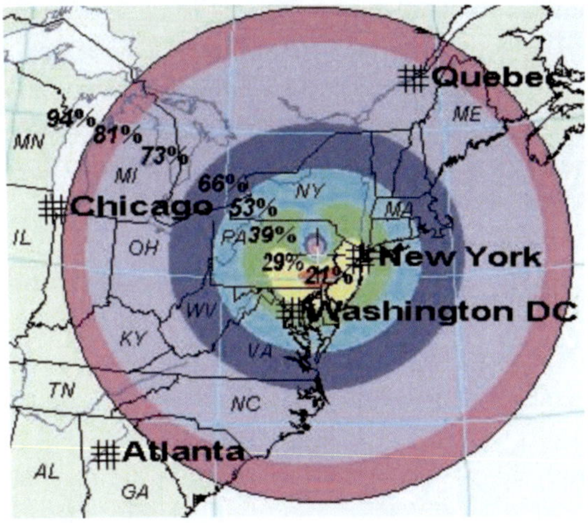

Figure 3-11. Percentage of Calls Completed 2 Days After EMP Event.

The illustrative results in **figure 3-12** highlight the value of operational GETS and WPS capabilities given that the call completion levels noted in **figure 3-12** would be unacceptable for NS/EP functions during the critical early stages of an emergency. The analysis performed as part of this EMP Commission effort did not explicitly examine the performance of these NS/EP services in an EMP attack. The call completion levels in **figure 3-12** would be seen as likely lower bounds for these services for the scenarios of interest examined.

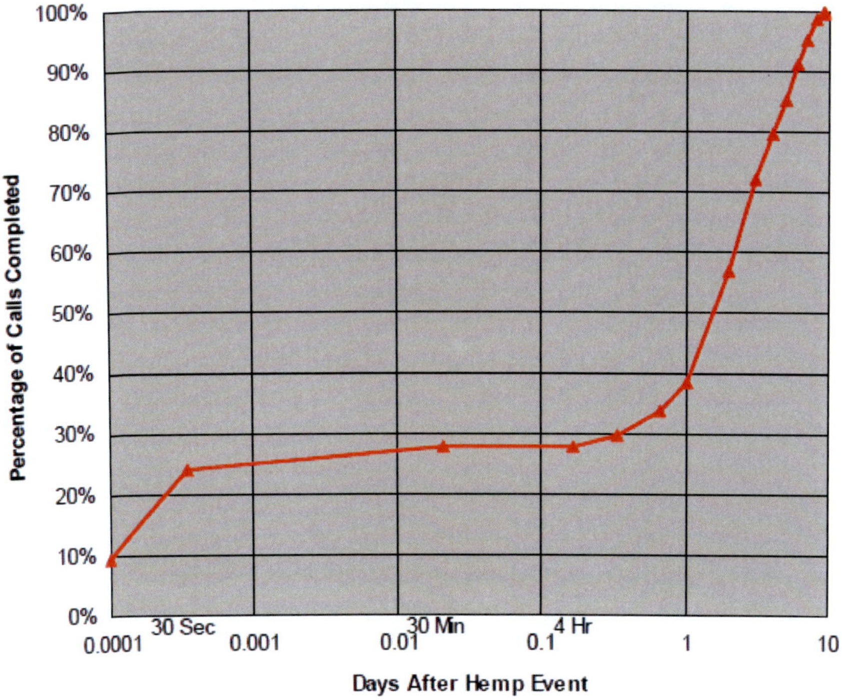

Figure 3-12. Percentage of Calls Completed at Time T (Logarithmic Time Scale) (Within EMP Contours).

The scenarios examined indicated that even in the case of minimal equipment damage, the functioning of NS/EP telecommunications services are critical to handling the spike in caller traffic expected to follow an EMP attack. This traffic tends to overwhelm the available telecom network capacity and results in degraded network performance. While operational experience exists with the current technologies that support NS/EP services, there is the need to make sure that NS/EP services operate effectively as new technologies, such as softswitches, are being introduced into the network. It is important to verify that this equipment will operate through an EMP attack under the stressful operating conditions that are anticipated in an emergency situation. The use of IP-related technology such as softswitches to support GETS and WPS services is at the initial point of deployment in local offices. Rigorous analysis is warranted prior to their major deployment to examine EMP survivability issues.

During sensitivity analysis, in the process of examining alternative cellular base station damage levels, an area of concern was identified within the cellular network system. Specifically, the area of concern was degradation of network performance due to EMP effects on critical databases, including the Home Location Registers (HLRs). HLRs contain key user information associated with cellular subscribers, such as account status and location. Within the wireless industry, the deployment approach to achieve HLR diversity (physical and geographic) is mixed. HLRs were not tested for susceptibility levels as part of the EMP Commission study, but using proxy numbers based on testing for circuit switching equipment, sensitivity studies show that it is possible to lose major calling areas in an EMP attack due to HLR degradation. In addition to EMP susceptibility testing, engineering polices and selective EMP hardening of these elements are options that should be examined in the future.

As noted in the Electric Power Section, loss of portions of the power grid is likely, even for a relatively low-level EMP attack. Our analysis indicates that, in a relatively low-level EMP attack, the direct impact on public telecommunications networks is likely to be dominated by the inability to handle ensuing spikes in call traffic. In such cases, the direct effects on equipment are expected to be largely transient and short term in nature (minutes to hours) with minimal manual restoration needed. However, should widespread loss of primary power occur, the survivability of the telecommunications network and associated NS/EP and other services will depend on the use of backup power capabilities and the rapidity with which primary power can be restored. Most public telecommunications equipment has a mix of battery, mobile generator, and fixed generator support if primary electric power is lost. A short-term loss of the electric power supporting most telecommunications networks today would not cause a major loss of telecom services. This is due to the existence of power backup systems and best practices supporting these critical systems that could sustain telecom services during short-term power outages.

The situation becomes more serious if the power outages are long term and widespread. In such cases, the likely loss of major telecommunications facilities would significantly reduce NS/EP services. A majority of residential telephones today depend on power from local central offices, which would be lost once the backup power at those offices is depleted. Other residential telephones also require commercial power to function. Thus, citizen ability to access 9-1-1 call centers would be a major concern in an extended power outage situation.

Hurricane Katrina in August 2005 damaged cell phone towers and radio antennas. The prolonged blackout resulting from Katrina exhausted the fuel supplies of backup generators servicing emergency communications. Consequently, emergency communications for police, emergency services, and rescue efforts failed. Significantly, these same nodes so critical to emergency communications—cell phone towers and radio antennas—are vulnerable to EMP attack. A protracted blackout resulting from an EMP attack would also exhaust fuel supplies for emergency generators, just as occurred during Hurricane Katrina.

Public telecommunications networks can successfully handle a local power outage or short-term outage, such as the August 14, 2003, Northeast blackout. However, a major concern exists with outage durations that range in weeks or months. The widespread collapse of the electric grid due to an EMP event would lead to cascading effects on interdependent infrastructures, as happened during the Katrina blackout. This may well lead to a long-term loss of telecommunications in extended geographic areas outside the power loss. This loss would cascade to any critical applications that depend on telecommunications. As such, telecommunications resilience would greatly benefit from steps to increase power grid and backup power reliability and availability time frames.

Telecommunications network managers have indicated that a key asset in any outage event is the ability to monitor the health of the network in real time to enable rapid response to identified problems. Given the increased level of automation in telecommunications networks coupled with reduction in personnel, it is critical that the telecommunications operations and control functions remain operational in an EMP event. In recovering from an EMP attack, telecommunications carriers will depend on hardware and software systems that help isolate problem areas and implement commands to initiate remediation efforts. Computer servers, personal computers, routers, and related equipment are key components that are housed in Network Management Centers. Carriers typically deploy the equipment in geographically diverse centers in which one center can back up the others. Effects to those centers are moderated in cases in which the centers are separated by distances larger than the EMP footprint.

Recommendations

Based on the analytical efforts performed by this Commission, the following steps are recommended to improve telecommunications performance during and after an EMP event:

- Successfully evolve critical NS/EP telecommunications services to incorporate the new technologies being embedded into telecommunications networks.
- Improve the ability of telecommunications services to function for extended periods without the availability of primary power.
- Adequately address infrastructure interdependency impacts in contingency planning.
- Identify critical applications that must survive an EMP event and address any shortfalls in telecommunications services that support these applications.

These recommendations are discussed in more detail in the next few sections.

Preventing Widespread Outages from New Technology

EMP is just one of the potential sources that would lead to stressing telecommunications networks. Understanding NS/EP service performance with respect to IP technology has benefits beyond application to EMP. This issue is in line with a U.S. government interagency Convergence Working Group (CWG) finding[11] that noted, "The FCC should task NRIC to assess the adequacy of interoperability testing between circuit and packet switch networks ... minimize the risk of feature interactions and the introduction of additional vulnerabilities affecting reliability, availability, and security of telecommunication services supporting NS/EP users."

High-profile network failures have occurred as new technologies were introduced into networks. Inadequate testing prior to widespread deployment has been highlighted as a major problem in lessons learned from past outages related to new technology introduction.[12] These offer an incentive for the testing of new technology supporting NS/EP services prior to widespread deployment of the technologies. The use of packet switching technology to support voice services such as GETS and WPS is at the initial point of deployment. Rigorous testing is warranted prior to major deployment. With early identification, specific system EMP vulnerabilities can be addressed prior to widespread deployment.

The following are specific steps to address technology introduction concerns:

- NCS[13] represents a logical organization to address these areas given its mission associated with the development and maintenance of NS/EP services. NCS should partner with other appropriate organizations to determine the effects of EMP on different types of telecommunications equipment, facilities, and operations by:

    - The testing and analysis of new technologies introduced into telecommunications networks that will support NS/EP services prior to widespread introduction into the public network. IP-related equipment should be a major near-term focus of this testing and analysis. This analysis should include examining the use of standards in terms of prevention and mitigation benefits.
    - Capturing the lessons learned from future outages associated with the expected growth of voice communications by nontraditional carriers and the tremendous growth in wireless communications. It is important that such lessons learned be captured in a systematic and fiscally prudent manner.

Historically, data captured by the Federal Communications Commission (FCC) on major outages has been extremely valuable in identifying and correcting problems as they are exhibited in deployed systems. Again, this is consistent with the EMP Commission's philosophy of preventing disastrous consequences from "cheap shot" attacks.

Reducing the Effects of Power Outages on the Telecommunications Infrastructure

In a power outage, telecommunications carriers typically depend on battery supplies that last from 4 to 8 hours and in some cases fixed and mobile generators that may have up to 72 hours of operating fuel. A key concern is the potential that major telecommunications facilities may not have primary power in the event of a long-term power outage of several weeks over a wide geographic area. Among the major concerns in such events are:

- The potential that major telecommunications facilities will not have prioritized access to fuel supplies on a long-term basis in the event of a long-term, wide-scale power outage.
- Facilities running on backup generators on a long-term basis will eventually require maintenance.

These concerns proved prescient when Hurricane Katrina struck in August 2005. Katrina caused a prolonged blackout that resulted in telecommunications failures precisely because of the above concerns regarding fuel supplies and maintenance for emergency generators.

After the August 2003 Northeast blackout, recommendations were put forward by the NRIC to help address this power dependency issue. As part of lessons learned discussed in an August 27, 2003, NRIC presentation on the impact of the 2003 Northeast blackout, telecom-specific references were made to re-evaluate the Telecommunications Electric Service Priority (TESP) program: "Power management and restoral practices at the tactical level are under review by carriers—may need modifications to the TESP program to mitigate additional risks," and "Development of TESP program for cellular networks to address priority restoration of critical cellular communications facilities is needed."[14] TESP promotes (on a voluntary basis) the inclusion of critical telecommunications facilities in electric service providers' priority restoration plans.[15]

Lessons learned from Katrina and the NRIC evaluation of the 2003 Northeast blackout form the underpinning for the following EMP Commission recommendations:

- Improve the ability of telecommunications to withstand the sustained loss of utility-supplied electric power:
  - Task the NCS and the North American Electric Reliability Corporation (NERC), or its successor, with providing, at a minimum, biannual status reports on the need for/adequacy of priority restoration of electric power by power utilities to selected telecommunications sites.
  - Task the Department of Energy (DOE) with exploring the adequacy of financial incentives to spur analysis of alternative powering sources that offer cost-effective and viable alternatives for telecom asset powering. For example, carriers are exploring new technologies such as fuel cells to support the powering of offices.

Adequately Addressing Interdependency Impacts in Contingency Planning

The potential impact of other interdependency effects, with a priority on NS/EP services, must be considered in any analysis of recovery planning. For example, the assumption of key personnel access to transportation to operations center sites or remote access to equipment should be addressed in contingency planning. With this in mind, the NCS would be a logical organization to address this area for critical national infrastructures. Specifically, the Commission recommends the following:

- Expand the role of the NCS within the Code of Federal Regulations (CFR) Part 215 (Federal Focal Point for EMP Information) to address infrastructure interdependencies related to NS/EP telecommunications services.

Supporting this recommendation is the need to exercise the National Response Framework to determine how well the plan addresses simultaneous degradation of multiple infrastructures. Industry personnel have suggested to the EMP Commission that a tabletop exercise considering this type of scenario would be extremely useful. Exercise results should be factored into the development of an EMP scenario to be included on the DHS list of National Planning Scenarios. Such an exercise would be invaluable in understanding the impacts of telecommunications failures on other infrastructure sectors and vice versa. Of particular concern is the impact of losing telecommunications on the operating effectiveness of Supervisory Control and Data Acquisition (SCADA) systems for infrastructures such as electric power and natural gas.

Specifically, the Commission recommends the following:

- Task DHS with developing exercises and an additional National Planning Scenario incorporating a large-scale degradation for multiple infrastructures over a wide geographic area as might occur in an EMP event.

Improving the Ability of Telecommunications Networks That Support Nationally Critical Applications to Survive EMP by Protecting Key Assets and Conducting Vulnerability Assessments

The Commission recommends the following:

- Task NCS to identify key telecommunications network assets whose degradation can result in the loss of service to a large number of users. These might include next-generation routing and transport equipment and wireless network elements such as HLRs and Visiting Location Registers (VLRs). Cellular base stations should be part of this analysis.
- Task NCS through DHS, in accordance with the CFR for Telecommunications Electromagnetic Disruptive Effects (TEDE) affecting NS/EP telecommunications, to work with government and multiple industries (e.g., Federal Reserve Board and BITS [financial services], Federal Energy Regulatory Commission [FERC] and NERC [electric power], and DHS and first responders [civilian restoration]) to determine whether a high-reliability telecommunications service or services supporting mission-critical applications is needed. If so, consider partial federal funding for this service.
- Establish a reporting process to be developed by the FCC, NCS, and the telecommunications industry for reporting major outages from wireless, data communications, and Internet carriers to the FCC, analogous to what is done for wireline carriers, thereby capturing lessons learned.

SECTION 4. BANKING AND FINANCE

Introduction

The financial services industry comprises a network of organizations and attendant systems that process instruments of monetary value in the form of deposits, funds transfers, savings, loans, and other financial transactions. Virtually all economic activity in the United States (U.S.) and other developed countries depends on the functioning of the financial services industry. National wealth is the sum of all economic value, as reflected in part in existing capital and financial transactions. Most simply, the financial services industry is the medium and record keeper for financial transactions and repository of national, organizational, and individual wealth.

Today, most significant financial transactions are performed and recorded electronically; however, the ability to carry out these transactions is highly dependent on other elements of the national infrastructure. According to the President's National Security Telecommunications Advisory Committee (NSTAC), "The financial services industry has evolved to a point where it would be impossible to operate without the efficiencies of information technology and networks."[16]

The automation of the financial services industry has spurred the growth of wealth by increasing greatly the amount of business that can be conducted on a daily basis. For example, "in the early 1970s, the New York Stock Exchange [NYSE] closed every Wednesday to clear backlogs from an average daily trading volume of 11 million shares."[17] Today, the Securities Industry Automation Corporation (SIAC) has no interruption in exchange operations and routinely handles an average daily trading volume of more than 3 billion shares.[18]

"SIAC is responsible for providing the highest quality, most reliable and cost-effective systems to support the current and future business needs of the New York Stock Exchange"[19] and other institutions. "SIAC's Shared Data Center alone is linked to the securities industry by more than a thousand communications lines over which an average of 70 billion bytes of data is transmitted daily."[20] SIAC's Secure Financial Transaction Infrastructure, "improves the overall resilience of the financial industry's data communications connectivity…and offers firms reliable access to… trading, clearing and settlement, market data distribution, and other services."[21]

The technological revolution has not been limited to giant corporations. The individual consumer has witnessed the growth of convenient, on-demand money-dispensing automated teller machines (ATM) in the United States from less than 14,000 in 1979[22] to more than 371,000 in 2003.[23]

The trend in the U.S. financial infrastructure is toward ever more sophisticated and powerful electronic systems capable of an ever increasing volume and velocity of business. The increasing dependence of the United States on an electronic economy, so beneficial to the management and creation of wealth, also increases U.S. vulnerability to an electromagnetic pulse (EMP) attack.

For example, the terrorist attacks of September 11, 2001, demonstrated the vulnerabilities arising from the significant interdependencies of the Nation's critical infrastructures. The attacks disrupted all critical infrastructures in New York City, including power, transportation, and telecommunications. Consequently, operations in key financial markets were interrupted, increasing liquidity risks for the U.S. financial system.[24]

An interagency paper jointly issued by the Office of the Comptroller of the Currency (OCC), the Federal Reserve Board (FRB), and the Securities and Exchange Commission (SEC), specifies clearing and settlement systems as the most critical business operations at risk for financial markets.[25] Because financial markets are highly interdependent, a wide-scale disruption of core clearing and settlement processes would have an immediate systemic effect on critical financial markets.[26]

Moreover, in December 2002, the FRB revised its policy and procedures for national security and emergency preparedness telecommunications programs administered by the National Communications System (NCS) to identify those functions supporting the Federal Reserve's national security mission to maintain national liquidity.[27] The FRB expanded the scope of services that would seriously affect continued financial operations if a telecommunications disruption of "a few minutes to one day" occurred.[28] These functions, which are listed below, "require same-day recovery and are critical to the operations and liquidity of banks and the stability of financial markets":[29]

- Large-value interbank funds transfer, securities transfer, or payment-related services
- Automated clearing house (ACH) operators
- Key clearing and settlement utilities
- Treasury automated auction and processing system
- Large-dollar participants of these systems and utilities.[30]

The increasing dependence of the United States on an electronic economy also adds to the adverse effects that would be produced by an EMP attack. The electronic technologies that are the foundation of the financial infrastructure are potentially vulnerable to EMP. These systems are also potentially vulnerable to EMP indirectly through other critical infrastructures, such as the electric power grid and telecommunications.

The Financial Services Industry

In a December 1997 study, *Financial Services Risk Assessment Report*, NSTAC described the financial services industry as comprising four sectors. This definition is reflected or shared in current U.S. government reports, regulations, and legislation that treat the financial services industry as having these components:

- Banks and other depository institutions
- Investment-related companies
- Industry utilities
- Third-party processors and other services.

Banks and Other Depository Institutions

In 2004, U.S. banks held more than $9 trillion[31] of domestic financial assets, and investment companies and other private institutions held about $17 trillion of the national wealth.[32] Banks and other depository institutions, including thrifts, credit unions, and savings and loan associations, are vital to the functioning of the economy. These institutions hold and provide access to deposits, provide loans, transfer funds, promote savings, and facilitate economic growth.

Commercial banks are the repository of the most financial assets of any depository institution. Commercial banks disseminate financial information, act as agents in buying and selling securities, serve as trustees for corporations or individuals, transfer funds, collect deposits, and provide credit. The top 10 commercial banks control nearly half of all assets held by banks.[33]

Credit unions, savings and loan associations, and savings banks generally are referred to as "other depository institutions." These institutions usually service households instead of businesses. Credit unions are the most financially significant of these institutions. By the end of 2004, credit unions had more than 85 million members and managed more than $668 billion in assets.[34]

The single most important banking institution is the Federal Reserve System. Established by the U.S. Congress in 1913, the Federal Reserve System is the central bank of the United States. This system does not deal directly with the general public, but with other banks. It is, in essence, the Nation's bank for commercial banks.

The primary purpose of the Federal Reserve System is to maintain the stability, safety, and flexibility of the financial system and contain systemic risk that may arise in the financial markets. The Federal Reserve accomplishes this mission by establishing monetary policy, by servicing financial institutions and other government agencies, and by regulating and supervising banks.

As the central bank of the United States, the Federal Reserve System extends emergency credit to commercial banks and controls interest rates, foreign exchange, and the money supply. The Federal Reserve also performs check-clearing and processing and transfer of government securities and funds between financial institutions.

Federal Reserve System banks are supervised by a Board of Governors who are appointed by the president and confirmed by the U.S. Senate; however, the banks are owned by private member banks. For administrative purposes, the United States is divided into 12 Federal Reserve Districts, each district served by a Federal Reserve Bank. The 12 Federal Reserve Banks are located in New York, Boston, Philadelphia, Richmond, Atlanta, Cleveland, Chicago, St. Louis, Kansas City, Dallas, Minneapolis, and San Francisco.

Investment-Related Companies

Unlike commercial banks, underwriters, brokerages, and mutual funds are not depository institutions. Rather, these institutions provide a wide range of services to institutional and individual investors. They act as intermediaries in pooling investments by a large group of customers and in market trades.

Investment banks and underwriters finance investments by government and commercial enterprises through stocks and bonds. Investment banks also arrange mergers. Currently, the largest 50 firms hold 90 percent of the market share.[35]

Brokerages help investors by acting as agents or intermediaries with commodities and securities markets. Brokerages advise clients, perform research, and place trades. "The securities brokerage industry in the United States includes fewer than 400 companies with combined annual revenue of over $100 billion. The top 50 companies hold over 80 percent of the market share."[36]

Mutual funds pool money from many people and institutions and invest it in stocks, bonds, or other securities. A portfolio manager is employed by the mutual fund to achieve its financial objective, such as providing a reliable source of investment income or maximizing long-term returns. The mutual fund market is dominated by 25 companies. The top five companies hold one-third of the market. The mutual fund industry holds about $8.1 trillion dollars in assets.[37]

Industry Utilities

Banks, including the Federal Reserve System, and investment-related companies, such as investment banks, brokerages, and mutual funds, all rely on industry utilities to transact business. Financial service utilities are the institutions that provide a common means for transferring, clearing, and settling funds, securities, and other financial instruments, as well as exchanging financial information.

Financial industry utilities have largely replaced paper transactions with electronic means. Check and cash transactions are still the largest number of financial transactions in the national economy. However, paper transactions are vastly surpassed in total value by electronic transactions through wire transfers, interbank payment systems, ACHs, and clearing and settlement systems for securities and other investments.

Modern financial services utilities have transformed the national economy from a paper system into an electronic system. Examples of some key industry utilities include FEDNET, Fedwire, ACH, Clearing House Interbank Payments System (CHIPS), the Society for Worldwide Interbank Financial Telecommunications (SWIFT), the National Association of Securities Dealers' Automated Quotation System (NASDAQ), the NYSE, the New York Mercantile Exchange (NYMEX), and the Depository Trust and Clearing Corporation (DTCC).

FEDNET is a communications system connecting all 12 Federal Reserve Banks nationwide and the financial services industry generally. FEDNET transfers funds in real time among banks and other depository institutions, performs real-time sales and record keeping for the transfer of government securities, and serves as ACH.

Fedwire is the primary national network for the transfer of funds between banks; the system currently serves approximately 7,500 institutions. Fedwire's book-entry securities transfer application allows banks and other depository institutions to transfer U.S. government securities. This network has enabled the Federal Reserve to largely replace paper U.S. government securities with electronic book entries. Transfers performed on Fedwire are irrevocable upon receipt and are settled immediately. The average value of a Fedwire funds transaction is about $3.9 million dollars.[38] In 2005, Fedwire processed an average daily volume of approximately 528,000 payments, with an average daily value of about $2.1 trillion.[39]

ACH was developed in the 1970s as an alternative to the traditional paper-based system for clearing checks. ACH electronic transactions include direct deposits of payrolls, pensions, benefits, and dividends and direct bill payments. The Federal Reserve annually processes about 36.7 billion ACH payments valued at $39.9 trillion dollars.[40]

CHIPS is an electronic system for interbank transfer and settlement. CHIPS is the primary clearing system for foreign exchange. "It processes over 285,000 payments a day with a gross value of $1.4 trillion." This includes 95 percent of all international U.S. dollar payments.[41]

The SWIFT provides stock exchanges, banks, brokers, and other institutions with a cost-effective, secure international payment message system. These messages are instructions between banks and other institutions regarding payments and transfers, not payments themselves. SWIFT carries approximately 8 million messages daily.[42]

The NASDAQ and the NYSE are the largest securities markets. NASDAQ is an electronic communications network that consolidates the quotations of multiple dealers, displayed in real time, and allows electronic trading. The NYSE offers similar electronic services. NASDAQ executed 957.9 million trades valued at more than $3.7 trillion dollars in 2004, and the NYSE traded a slightly lesser amount.[43]

The NYMEX trades on futures contracts such as unleaded gasoline, heating oil, crude oil, natural gas, and platinum. NYMEX typically conducts crude oil transactions involving the total daily production of the entire world.

The DTCC settles securities trades for participant banks and is the largest securities depositary in the world. In 2004, the company completed financial settlement for a quadrillion dollars in securities transactions. DTCC keeps records on securities and conducts transactions electronically. Annually, DTCC participants deliver securities valued at about $4.5 trillion to DTCC to make electronic records of ownership.[44]

Third-Party Processors and Other Services

Third-party processing companies are technology companies that provide electronic processing services to financial institutions. Banks and other financial institutions can cut overhead by contracting with third parties to perform the mechanics of electronic transactions. Technology-related outsourcing is especially appealing because of dynamic changes in technology. The high cost and complexity of new technologies has driven many banks into partnerships with third-party specialists in the field of electronic finance. Services typically offered by third-party processors include data center management, network management, application development, check and statement processing, mutual fund account processing, and electronic funds transfer.

Vulnerability to EMP

The financial infrastructure is highly dependent on electronic systems, which should be clear from the preceding discussion. Virtually all transactions involving banks and other financial institutions happen electronically. Virtually all record keeping of financial transactions are stored electronically. Just as paper money has replaced precious metals, so an electronic economy has replaced the paper one. The financial infrastructure is a network of simple and complex electronic machinery, ranging from telephones to mainframe computers, from ATMs to vast data storage systems.

The electronic technologies that are the foundation of the financial infrastructure are potentially vulnerable to EMP. These systems also are potentially vulnerable to EMP indirectly through other critical infrastructures, such as the power grid and telecommunications.

The financial services industry and knowledgeable experts on the security of that industry judge that the industry is highly robust against a wide range of threats. The NSTAC, for example, notes that the leading financial institutions take a multilayered approach to building robustness and recoverability into their systems:

> *Operational data centers are engineered from the ground up with survivability in mind. Some are hardened with thick concrete walls and protected with extensive perimeter security measures equivalent to military command posts. Most have uninterruptible power supplies, generators, and on-site fuel storage sufficient to allow the facility to run independently of the power grid ranging from a few hours to over a month. External telecommunications links are diversely homed, with multiple building access points and connections to more than one central office...wherever possible. Operational procedures within the data center are designed to minimize the risk of human errors causing interruptions, and most or all data files are copied and stored on disk or tape at off-site facilities*[45]

NSTAC also observes that, "Numerous natural and man-made disasters...have forced financial institutions to test and refine their disaster recovery capabilities."[46] The financial services industry's dependence on other infrastructures has been tested in real emergencies. For example, in 1988, a fire in the Ameritech central office in Hinsdale, Illinois, disabled long-distance telecommunications for the Chicago Board of Trade and other major institutions. Wall Street was blacked out for nearly a week by an electrical fire in a Consolidated Edison office in August 1990. In April 1992, underground flooding in Chicago caused sustained telecommunication and power outages. Financial institutions faced widespread electrical power outages in the West during the summer of 1996 and in the Northeast during the summer of 2003.

"In addition," according to NSTAC, "the industry weathered one of the worst terrorist attacks in recent history":

> *The World Trade Center bombing on February 26, 1993, struck at the industry's heart, affecting the New York Mercantile Exchange and many securities dealers and otherwise disrupting activities throughout Wall Street. Numerous problems with facilities, systems, procedures, and staffs were encountered as firms scurried to recover, and some securities firms' operations were shut down temporarily. However, none of the most critical services were affected, and the effect on the economy as a whole was minimal.*[47]

The financial services industry also weathered the more devastating terrorist attack on September 11, 2001, that destroyed the World Trade Center. NSTAC found that these types of events, "led to improved robustness of the financial services infrastructure."

NSTAC's judgment that the financial services industry enjoys robust survivability against a wide range of threats is seconded by the National Academy of Sciences (NAS) in its study, *Making the Nation Safer: The Role of Science and Technology in Countering Terrorism* (2002). According to the NAS, the U.S. financial infrastructure is highly secure because of the redundancy of its electronic systems: "While no law of physics prevents the simultaneous destruction of all data backups and backup facilities in all locations, such an attack would be highly complex and difficult to execute, and is thus implausible."[48]

However, the NSTAC and NAS studies were focused primarily on the threat to the financial services industry from cyberterrorists using computer-based attacks. These studies did not evaluate the threat from EMP attack.

An EMP attack would pose the very kind of simultaneous and widespread threat postulated by the NAS that would be fatal to the financial infrastructure but judged by them to be too difficult to execute and implausible for cyberterrorists. EMP effects propagate at the speed of light and would cover a broad geographic area. Such an attack potentially could achieve the NAS criteria for financial infrastructure catastrophe: "simultaneous destruction of all data backups and backup facilities in all locations."[49]

An EMP would probably not erase data stored on magnetic tape. However, by shutting down power grids and damaging or disrupting data retrieval systems, EMP could deny access to essential records stored on tapes and compact discs (CD). Moreover, because EMP physically destroys electronic systems, it is also in the category of threats that NSTAC concludes are more worrisome than cyberterrorism: "Physical attacks remain the larger risk for the industry."

The vast majority of electronic systems supporting the financial infrastructure have never been tested, let alone hardened, against EMP. Yet the enormous volume, speed, and accuracy required of the electronic infrastructure supporting the financial services industry allow little or no room for error. Financial operations could not tolerate the kind of disruptions or mass systemic destruction likely to follow an EMP attack.

For example, CHIPS interbank transactions typically involve about $1.4 trillion dollars of business every day, or some $182 billion dollars every hour.[50] CHIPS and Fedwire routinely receive 5 to 10 funds transfer messages each second during peak traffic periods.[51] The Options Clearing Corporation manages $1.05 billion in average daily premium settlements.[52] On Christmas Eve 2004, a single credit card association processed over 5,000 transactions per second.[53] Financial institutions also must store tremendous amounts of data. Terabyte portfolios (containing 1 trillion bytes) are now common, and some databases exceed a petabyte (1,000 trillion bytes). Changes in these huge databases must be recorded at the end of every business day.

"Dealing with this kind of volume, industry utilities cannot afford any interruption in service," according to NSTAC. An EMP attack, with its potential to disrupt communications possibly for days, weeks, or months and to destroy or change databases, would place the financial infrastructure at risk.

Although the financial services industry has survived and learned from natural and man-made disasters, those disasters also have exposed vulnerabilities that could be exploited by an EMP attack. According to the staff director for management of the FRB, the terrorist attack of September 11, 2001, on the World Trade Center exposed telecommunications and the concentration of key facilities as serious weaknesses of the financial services industry. Equity markets closed for 4 days, until September 15, due to failed telecommunications. The NYSE could not reopen because key central offices were destroyed or damaged, leaving them unable to support operations. According to this senior government official, Fedwire, CHIPS, and SWIFT would cease operation if telecommunications were disrupted. He further observed that ACH, ATMs, and credit and debit cards all depend on telecommunications. Disruption of these systems would force consumers to revert to a cash economy.[54]

Further, response to the Northeast power outage in August 2003 has been depicted as a triumph for the financial services industry safeguards implemented since the terrorist attacks of September 11, 2001. But this is not the whole picture. Some analysts observe that the blackout happened under nearly ideal conditions to facilitate financial industry recovery. The blackout happened on a Thursday at 4:10 p.m., after the 4:00 p.m. closing time for financial markets, and it was largely over for the financial industry by 9:00 a.m. the following Friday morning. Business was also light, at its nadir, as is usual during August.

Even so, recovery from the 2003 blackout still required many in the financial industry to work overnight. The American Stock Exchange did not open because its air conditioners would not operate. Many traders could not get to work on Friday because the transportation system was paralyzed. Some companies were unable to reach the NASDAQ electronic exchange by telephone. Many ATMs failed. Many of the 1,667 banks in New York City closed on Friday because of continuing power outage. Many industries with back-up generators, like KeyCorp in Cleveland, were unprepared for a blackout that lasted for more than a few hours, and they had difficulty getting diesel fuel.

The fortunate timing and short duration of the 2003 blackout affected the financial industry for a relatively brief period. Nonetheless, banks had to compensate for financial imbalances by borrowing $785 million dollars from the Federal Reserve System. This was 100 times the amount borrowed the previous week, and the greatest amount borrowed since the week after the September 11 attacks.[55] Most economists concur that the blackout had a small but measurable effect on the U.S. third-quarter economic growth.

These observations suggest that, if an EMP attack were to disrupt the financial industry for days, weeks, or months rather than hours, the economic impact would be catastrophic. The prolonged blackout resulting from Hurricane Katrina in August 2005 is a far better example than the Northeast blackout of 2003 of the challenge that would be posed to the financial infrastructure from EMP. The Katrina blackout, comparable to a small EMP attack, disrupted normal business life for months and resulted in a staggering economic loss that is still an enormous drain on the national economy.

The financial network is highly dependent on power and telecommunications for normal operations. Widespread power outages would shut down the network, and all financial activity would cease until power was restored, as happened during Hurricane Katrina. Even if power were unaffected or restored in short order, full telecommunications are required to fully enable the financial network. If critical elements within the telecommunications infrastructure were negatively affected by the EMP attack (i.e., at main and local switches), the financial network would be impacted negatively to some degree and consequently be highly dependent on the telecommunication recovery timelines before it could be brought back online with the required capability and capacity.

The extent to which the financial network is able to function as it is being brought back online will be highly dependent on the level of damage incurred by the network as a result of the EMP attack.

Consequences of Financial Infrastructure Failure

Despite the robustness of U.S. financial infrastructures against a wide range of threats, they were not designed to withstand an EMP attack. Indeed, the highly sophisticated electronic technologies that make the modern U.S. financial infrastructure possible are the components most vulnerable to EMP.

An EMP attack that disrupts the financial services industry would, in effect, stop the operation of the U.S. economy. Business transactions that create wealth and jobs could not be performed. Loans for corporate capitalization and for private purposes, such as buying homes and automobiles could not be made. Wealth, recorded electronically in bank databases, could become inaccessible overnight. Credit, debit, and ATM cards would be useless. Even reversion to a cash economy might be difficult in the absence of electronic records that are the basis of cash withdrawals from banks. Most people keep their wealth in banks and have little cash on hand at home. The alternative to a disrupted electronic economy may not be reversion to a 19th century cash economy, but reversion to an earlier economy based on barter.

In the immediate aftermath of an EMP attack, banks would find it very difficult to operate and provide the public with the liquidity they require to survive; that is, to buy food, water, gas, or other essential supplies and services. Modern banking depends almost entirely on electronic data storage and retrieval systems for record keeping and to perform account transactions. An EMP attack that damages the power grid or electronic data retrieval systems would render banking transactions virtually impossible as a practical or legal matter.

Operating a banking system using paper and handwritten transactions would be difficult without access to the information contained in electronic records. If a makeshift paper banking system could be organized on an emergency basis, such a system would be fraught with the risk of fraud, theft, and costly mistakes. Such a system would not be consistent with the cautious behavior and natural interest of banks in assigning highest priority to protecting financial assets. Protocols and business standards that are required of banks under their charters for insurance purposes and to protect them from legal liability assume the existence of modern electronic banking systems and the reliability, redundancy, and surety that such systems provide.

A survey by Commission staff of natural and man-made disasters found no case in which banks, bereft of their electronic systems because of blackout, reopened their doors and did business by hand. Unless banks have well-prepared contingency plans in place to revert to paper and handwritten transactions in advance of a crisis, it is very doubtful that bank managers would have the capability, authority, or motivation to attempt a paper and handwritten banking system in the aftermath of an EMP attack. Unless directed by federal authority to create contingency plans for operating without electricity, it is doubtful the business community would undertake such plans on its own.

In the aftermath of an EMP attack, individuals and corporations would have many sound reasons for being cautious, risk averse, and unwilling to resume business as usual. Once power, telecommunications, and transportation are restored, even if restored promptly, within a matter of days, psychological concerns that affect economic revitalization may linger. Full recovery will require restoring the trust and confidence of the business community in the infrastructures, in financial institutions, and in the future. The Great Depression outlasted its proximate causes by many years, despite strenuous efforts by the Federal Government to implement financial reforms and jump-start the economy, in part because businesses were unwilling to risk their capital in a system that had lost their confidence.

The Department of the Treasury and the SEC share the view that failure of electronic systems supporting the critical infrastructure for even one business day threatens the financial system with wide-scale disruption and risk to one or more critical markets. Indeed, the *Interagency Paper on Sound Practices to Strengthen the Resilience of the U.S. Financial System*, by the Department of the Treasury and the SEC advocates "the overall goal of achieving recovery and resumption within two hours after an event." It states:

> *In light of the large volume and value of transactions/payments that are cleared and settled on a daily basis, failure to complete the clearing and settlement of pending transactions within the business day could create systemic liquidity dislocations, as well as exacerbate credit and market risk for critical markets. Therefore, core clearing and settlement organizations should develop the capacity to recover and resume clearing and settlement activities within the business day on which the disruption occurs with the overall goal of achieving recovery and resumption within two hours after an event.*[56]

Partial or small-scale disruption of the financial infrastructure would probably be enough to bring about a major economic crisis. Nonfunctioning ATM machines, for example, and other impediments to obtaining cash might well undermine consumer confidence in the banking system and cause a panic. NSTAC observes that the ultimate purpose behind all the financial industry's security efforts is to retain consumer confidence: "The ability of an institution to maintain the trust, and hence, the business, of its customers is viewed as an even greater value than the dollars and cents involved."[57] A related NAS study concludes that an attack that destroys only electronic records would be "catastrophic and irreversible."[58] Although it is highly unlikely that stored financial data on magnetic media would be damaged by EMP, the electronic systems for retrieving data are potentially vulnerable to EMP and are dependent on a vulnerable power grid. Data and essential records are useless if inaccessible. According to the NAS, "Irrecoverable loss of critical operating data and essential records on a large scale would likely result in catastrophic and irreversible damage to U.S. society."[59]

Recommendations

Securing the financial services industry from the EMP threat and from other threats is vital to the national security of the United States. The Federal Government must ensure that this system can survive sufficiently to preclude serious, long-term consequences.

The Department of Homeland Security, the FRB, and the Department of the Treasury, in cooperation with other relevant agencies, must develop contingency plans to survive and recover key financial systems promptly from an EMP attack.

Key financial services include the means and resources that provide the general population with cash, credit, and other liquidity required to buy essential goods and services. It is essential to protect the Nation's financial networks, banking records, and data retrieval systems that support cash, check, credit, debit, and other transactions through judicious balance of hardening, redundancy, and contingency plans.

The Federal Government must work with the private sector to ensure the protection and effective recovery of essential financial records and services infrastructure systems from all deliberate adverse events, including EMP attack. Implementation of the recommendations made by the Department of the Treasury, the FRB, and the SEC in their *Interagency Paper on Sound Practices to Strengthen the Resilience of the U.S. Financial System* to meet sabotage and cyberthreats that could engender requirements for protection and recovery should be expanded to include expeditious recovery from EMP attack as follows:

- "Every organization in the financial services industry should identify all clearing and settlement activities in each critical financial market in which it is a core clearing and settlement organization or plays a significant role" that could be threatened by EMP attack.
- Industry should "determine appropriate recovery and resumption objectives for clearing and settlement activities in support of critical markets" following an EMP attack.
- Industry should be prepared to cope with an EMP attack by maintaining "sufficient geographically dispersed resources to meet recovery and resumption objectives.... Back-up sites should not rely on the same infrastructure components (e.g., transportation, telecommunications, water supply, electric power) used by the primary site. Moreover, the operation of such sites should not be impaired by a wide-scale evacuation at or inaccessibility of staff that service the primary site."
- Industry should "routinely use or test recovery and resumption arrangements.... It is critical for firms to test back-up facilities of markets, core clearing and settlement organizations, and third-party service providers to ensure connectivity, capacity, and the integrity of data transmission" against an EMP attack.[60]

SECTION 5. PETROLEUM AND NATURAL GAS

Introduction

The United States economy is dependent on the availability of energy. While much of that energy originates in natural resources of coal, hydroelectric, and nuclear materials and is distributed to users through the electric power grid, more than 60 percent of all U.S. domestic energy[61] usage derives from petroleum (about 40 percent) and natural gas (more than 20 percent) and is distributed to users through an extensive national pipeline system. Refined petroleum products and natural gas power our cars, heat our homes, energize our factories, and comprise critical elements of industrial materials ranging from fertilizers to plastics, all enabling the normal functioning of our energy intensive civil society. In 2006, according to the Annual Energy Review, the United States imported an average of 10 million barrels of crude oil and 11.5 billion cubic feet of natural gas every day. Domestically the United States produced about 5 million barrels of crude and 50.6 billion cubic feet of dry gas daily. All of these energy resources were delivered from their points of production or ports of entry to users or further distribution points through the national pipeline system.

While the closely related petroleum and natural gas infrastructures comprise a variety of production, processing, storage, and delivery elements, as described in the next section, the focus of this Section will be on the delivery system. In particular, we shall focus on the potential electromagnetic pulse (EMP) vulnerability of the more than 180,000 miles of interstate natural gas pipelines and the more than 55,000 miles of large — 8-inch to 24-inch diameter — oil pipelines.[62] We shall point to the potential vulnerabilities of the electronic control systems — supervisory control and data acquisition systems (SCADA) — that were discussed in general terms in Section 1, but whose criticality and centrality for the operation of the petroleum and natural gas infrastructure distribution systems are particularly prominent. Control system components with low voltage and current requirements, such as integrated circuits, digital computers, and digital circuitry, are ubiquitous in the U.S. commercial petroleum and natural gas infrastructures, and EMP-caused failures can induce dangerous system malfunctions resulting in fires or explosions.

Infrastructure Description

Petroleum

The petroleum infrastructure can be divided into two parts: the upstream sector, which includes exploration and production of crude oil, and the downstream sector, which comprises the refining, transmission, and distribution of the finished petroleum product.

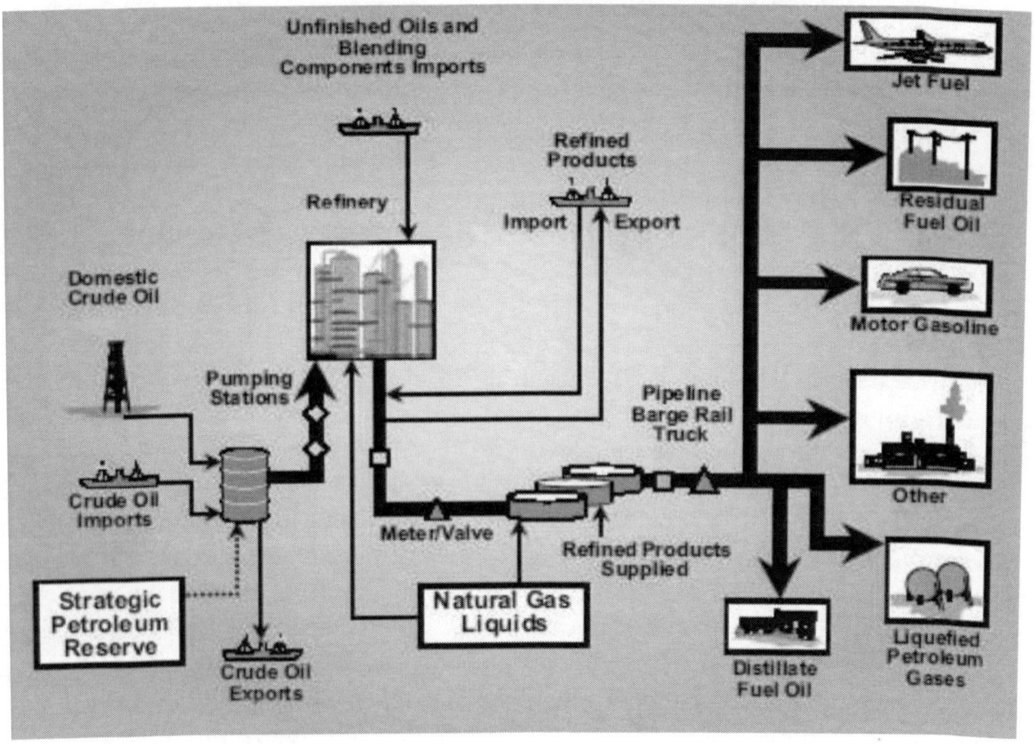

Figure 5-1. Petroleum Infrastructure.[63]

Physical components of the upstream sector include land oil wells and waterborne oil rigs for exploration, drilling, and extraction of crude oil. In 2006, there were 274 rotary rigs operating on- and off-shore in the United States and 501,000 crude oil producing wells (**figure 5-1**). In addition, many elements of the production of crude oil are located abroad, because the majority of U.S. oil is imported.

In contrast to the production stages of petroleum, the United States is the largest producer of refined petroleum products in the world. In 2006, 149 refineries were producing approximately 23 percent of the world's refinery output. These refineries range in production capabilities from 5,000 barrels to approximately 500,000 barrels per day. Nearly one-half of America's refining capacity is located along the Gulf Coast, mostly in Texas and Louisiana. Other major refineries are found throughout the Midwest and in California, Washington, and along the East Coast of the United States.

The most pervasive physical element of the oil infrastructure is the extensive transmission network that moves crude oil from the field to the refineries for processing and brings the finished products to the consumer. Pipelines are the safest and most economical way to accomplish this and account for nearly 50 percent of all crude oil received in domestic refineries in 2006. Tankers transport an additional 46 percent of the crude oil received by refineries, with the remaining crude oil delivered to refineries by barge, rail tank car, and truck. There are approximately 55,000 miles of crude oil trunk lines (8-inch to 24-inch diameter) and an additional 30,000 to 40,000 miles of smaller gathering lines (2-inch to 6-inch diameter) across the United States. The trunk lines connect regional markets, while the smaller gathering lines transport crude oil from the well — on- or off- shore — to larger trunk lines and are located mainly in Texas and Louisiana. Movement of the refined products, such as gasoline, diesel, and jet fuel, to the marketplace is done largely by tankers. In addition, there are approximately 95,000 refined product pipelines nationwide, varying in diameter from 8 to 12 inches to 42 inches, that bring products to their final destinations.

Storage facilities are an integral part of the movement of oil by rail, highway, pipeline, barge, and tanker and can be aboveground, underground, or offshore. In the United States, the most common storage tank is aboveground and made of steel plates. Most underground storage tanks are made out of steel as well. These storage facilities are located at each node in the production and distribution of petroleum and include tanks at the production field, marine terminals, refineries, pipeline pumping stations, retail facilities, car gasoline tanks, and home heating tanks.

In 2006, the United States imported about 60 percent of its petroleum consumption from abroad. Four thousand U.S. off-shore platforms, 2,000 petroleum terminals, and 4,000 oil tankers belonging to the world's energy trading nations and unloading petroleum at 185 ports in the United States, must also be counted as part of the petroleum infrastructure.

Natural Gas

The natural gas infrastructure comprises production wells, processing stations, storage facilities, and the national pipeline system (see **figure 5-2**).

In 2006, there were 448,461 gas- and condensate-producing wells[64] distributed among 63,353[65] oil and gas fields in the United States. There were more than 500 natural gas processing plants[66] and more than 1,400 compressor stations that maintain pressures in the pipeline and assure the forward motion of the transmitted gas supply. Storage facilities included 394 active underground storage fields, consisting of depleted oil and gas fields, aquifers, and salt caverns, five liquefied natural gas (LNG) import facilities, and 100 LNG peaking facilities. The pipeline system consists of more than 300,000 miles of inter-state and intrastate transmission lines and an additional 1.8 million miles of smaller distribution lines that move gas closer to cities and to individual homes and business.

Most of the natural gas consumed in the United States is produced domestically. Historically domestic production has accounted for around 85 percent of U.S. consumption with imports from Canada making up the remaining 15 percent. In recent years, domestic production has fallen to about 75 percent of consumption with the remainder imported from Canada. In 2005, five states — Texas, Oklahoma, Wyoming, Louisiana, and New Mexico — accounted for 77 percent of domestic natural gas production.

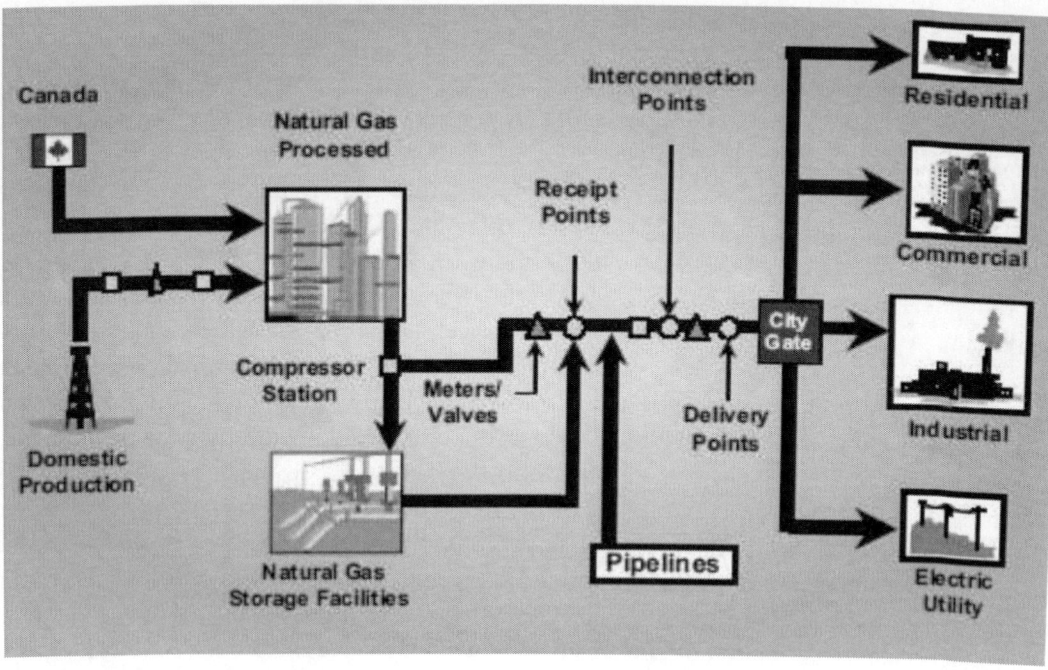

Figure 5-2. Natural Gas Infrastructure.

Direct Effects of EMP on Petroleum and Natural Gas Infrastructure

The infrastructure described in the previous section is dependent on the continuous operation of a wide variety of electrical components: pumps to extract fuel from wells and manage its movement through pipelines, electrically driven systems to process materials in refineries, transportation systems to deliver fuels to users from storage sites, point-of-sale electronics to process transactions to retail customers, and so on — all of which represent potential points of vulnerability to an EMP pulse. We shall focus here on the vulnerability due to only one of these components — SCADA — because they represent a ubiquitous presence across all the different infrastructure elements and play a series of critical roles whose loss would severely compromise, or in some instances eliminate altogether, the ability of the infrastructure to function.

SCADA systems themselves, and their tested vulnerability to electromagnetic pulses, were described in some detail in Section 1, the introductory Section to this volume, and we shall not repeat that here. Instead we describe the particular role of SCADAs within the petroleum and natural gas infrastructure, and then consider the consequences of an event which degrades or destroys the control and monitoring functions performed by the SCADAs.

Petroleum Infrastructure and SCADA

SCADAs play a critical role at every stage of the oil industry's life cycle: production, refining, transportation, and distribution. Automation within the oil industry begins at the resource exploration stage and ends with final delivery to the customer. At each step, process control and SCADA are used not only to ensure that operations are efficient, but also that strict safety measures are maintained to prevent injuries and fatalities, fires and explosions, and ecological disasters.

SCADA systems, for example, are deployed in production fields, pipeline gathering systems, and along pipelines to monitor and adjust various operating parameters. These monitoring functions assist oil companies in preventing leaks and other hazardous conditions, as well as minimizing the impact of those that do occur.

These systems, which involve two-way traffic requiring paired channels, allow a master station to monitor and control the status of a multitude of measurements and tolerance limits at wellheads, pump stations, and valves, thus eliminating the need for constant manual surveillance. **Figure 5-3** presents a typical SCADA system for offshore oil production and onshore oil distribution, showing the use of remote terminal units (RTUs) and distributed control systems (DCS) at remote locations and their connection with the master terminal units (MTUs) through various communication media.

Pumping facilities that produce thousands of horsepower of energy and metering facilities that measure thousands of barrels per hour are routinely operated remotely via these SCADA systems. They can be properly operated only by using extremely reliable communications systems. The control aspect may include controls to a well pump to increase or decrease output or to shut down altogether. Pipeline controls may include changing routing, increasing or reducing the flow of the liquids or gases, and other functions. However, some pipeline facilities still require manual operation.

Process control is concerned with maintaining process variables, temperatures, pressures, flows, compositions, and the like at some desired operating value. Process control systems within refineries, along pipelines, and in producing fields were previously closed and proprietary. These control processes are now moving toward open architecture and commercially available software. The oil infrastructure now relies on e-commerce, commodity trading, business-to-business systems, electronic bulletin boards, computer networks, and other critical business systems to operate and connect the infrastructure. These assessment and control tools depend to a large degree on telecommunications and associated information technologies. Telecommunication in this context refers to a system of information linkages and data exchanges that include SCADA, the associated SCADA communication links, control systems, and integrated management information systems.

Natural Gas Infrastructure and SCADA

SCADA is essential to modern natural gas operations. These systems provide the near-real-time data flows needed to operate efficiently in a deregulated environment. In addition, SCADA provides reporting of all transactions, establishing financial audit trails.

The key to effectively managing natural gas deliveries to customers is knowing what is happening along an interstate or intrastate pipeline system at all times. This is accomplished with Gas Control — a centralized command post that continuously receives information from facilities along the pipeline and disseminates information and operational orders to equipment and personnel in the field (see **figure 5-4**).

Through the use of SCADA equipment, Gas Control monitors volumes, pressures, and temperatures, as well as the operating status of pipeline facilities. Using microwave, telephone, or communication satellites, SCADA provides the Gas Control operator with information on the volume of natural gas flowing into the system and the volume of gas delivered to customers and gives the ability to quickly identify and react to equipment malfunctions or incidents. SCADA also gives Gas Control the capability to remotely start or stop compressors or open or close valves, thereby varying flow volumes to meet changes in customer demand for natural gas. Before the advent of SCADAs, all such functions, including tedious flow computations, were performed manually.

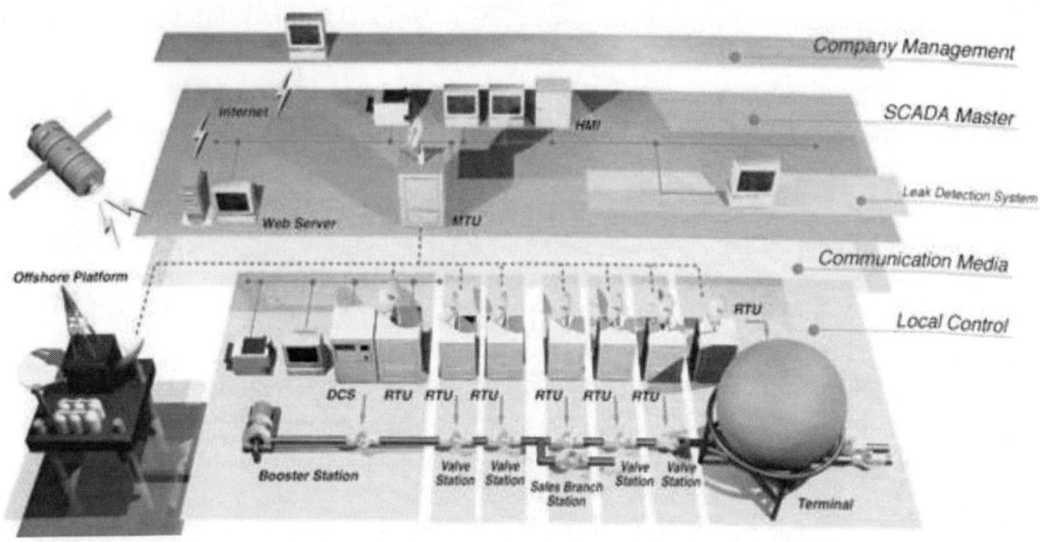

Figure 5-3. Typical SCADA Arrangement for Oil Operations.

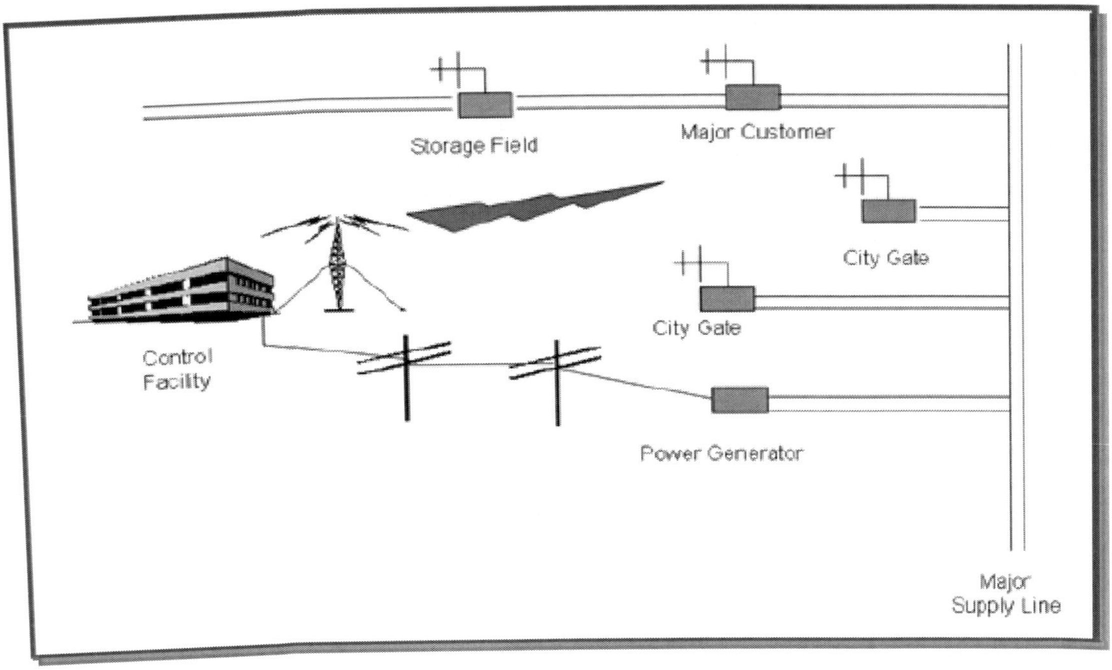

Figure 5-4. SCADA Integrates Control of Remote Natural Gas Facilities.

Automation of natural gas operations employs electronic components and technology to a high degree. Many of these components use simple mechanical or electrical properties to perform their defined roles, but an increasing number of them are computer-based. The major components and subsystems are RTUs, programmable logic controllers (PLC), MTUs, and communication systems, both wired and wireless. The total SCADA structure also includes control centers, information technology, personal computers (PC), and other peripheral technologies. RTUs and PLCs are usually located at the remote operational sites and connected to the MTUs and communication infrastructure through the communications network.

Effects of an EMP Event on the U.S. Petroleum and Natural Gas Infrastructures

There are few empirical data to support definitive statements regarding the precise effects of an EMP event, should one occur. We can only extrapolate from what is known about the effects of various levels of EMP testing and what is indicated by other types of ongoing tests. It is evident that electronic devices, particularly those incorporating solid-state circuitry are, to varying degrees, susceptible to the effects of an EMP event.

The principal electronic components of a SCADA system, those devices most vulnerable to an EMP attack, are found in all the major subsystems of the SCADA installation. The MTU is a modern computer, with various solid-state circuits embedded on the microchips contained inside. An EMP event may affect these, either as a temporary disruption, which, if not automatically rebooted, might require manual intervention, or with permanent damage. If MTUs are not physically damaged, it may not be obvious whether their functional state has altered. As discussed earlier, loss of the MTU would blind the Control Center personnel to system data and performance. The physical system (e.g., pipelines, refineries) would continue to operate within the limits of the preprogrammed RTU controls, assuming that these components also have not been adversely affected by the EMP event.

The RTUs and PLCs used in today's SCADA systems rely on solid-state circuits to maintain their programming and to carry out the directives issued through those pro- grams. This design makes the RTU and PLC inherently vulnerable to an EMP event. Although small, remote installations potentially have less exposure, it can be assumed that some or all of the RTUs and PLCs would be affected by an EMP event. As in the case of the MTU, affected embedded, integrated chips are suspect, even if the damage is not total and perhaps not immediately evident.

Gas

Functional loss of the RTU and PLC results in loss of supervisory control at that location. The equipment is unable to direct changes in pressure to match changes in demand requirements for the natural gas sector. The gas delivery system should continue to operate, and natural gas should continue to flow, but ultimately the system may reach extreme conditions. Due to the presence of backup emergency pressure regulation, it is unlikely that such a failure would lead to an unsafe condition, one that would cause a rupture or explosion. The most likely result, given no manual intervention, would be significant loss of pressure after some period of time, leading to massive service disruption.

Currently, if any component of the control system (e.g., RTU, PLC, MTU) for the natural gas infrastructure fails, the system still has the mechanical ability to operate as it did in the days before SCADA. An EMP-induced false signal might affect operation if the signal unexpectedly closed a valve instead of keeping it open. The SCADA system would then have no ability to adjust to changing conditions; however, except in extreme cases such as peak winter demand conditions, it should be able to maintain deliveries until field personnel arrive and institute manual control. Discussions with natural gas system operators provide a consensus that it would be highly unlikely that the natural gas pipe- line system would be shut down immediately if it is recognized that there is problem with the field data.

Oil

If the SCADA system for an oil pipeline is inoperative due to the effects of an EMP event, it is the opinion of a number of former pipeline personnel that operations would have to be shut down. A petroleum pipeline failure can be catastrophic. Leaking oil could contaminate water supplies and cause disastrous fires. Based on their experience, it has been stated that companies that operate any type of complex pipeline system today do not have enough personnel to manually operate the system using on-site operators with telephone communications (which may not be available after an EMP event) to a central control center, due in part to the multiple sites that need to be monitored and controlled during an emergency. Over the past decade, there has been a trend to increase remote control capability while reducing personnel in the oil and natural gas pipeline industry.

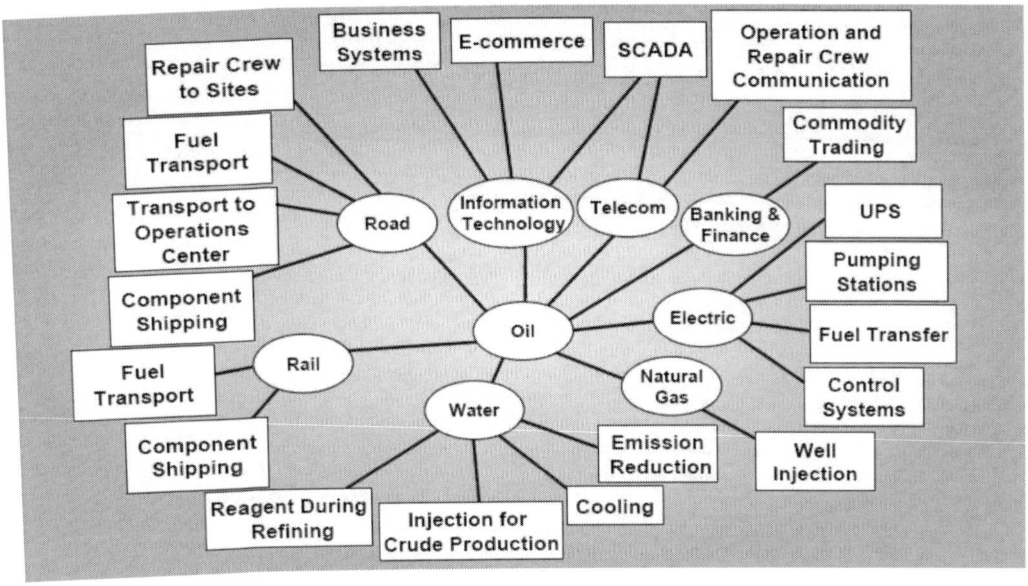

Figure 5-5. Examples of Oil Interdependencies.

U.S. refineries are critically dependent on the computers and integrated circuitry associated with process control, which are vulnerable to EMP effects. Discussions with plant managers and process control engineers at a number of refineries gave a nearly unanimous response that loss of process control would lead to refinery shutdown. A number of refineries stated they maintain an emergency override fail-safe system that institutes a controlled shutdown of the refinery if various SCADA parameters are out of range. However, the very short notice of a process control outage and the emergency shutdown procedure a refinery must undergo significantly increase the potential for equipment damage and lost production.

Indirect Effects of EMP: Accounting for Infrastructure Interdependencies

Infrastructure interdependency was discussed from a more general perspective in Section 1. The petroleum and natural gas infrastructures provide illustrative examples of such interdependencies as illustrated in figures 5-5 and 5-6.[67]

The petroleum and natural gas infrastructures are critically dependent on the availability of assured electric power from the national grid, as well as all the other critical national infrastructures, including food and emergency services that sustain the personnel manning these infrastructures. In turn, all these infrastructures rely on the availability of fuels provided by the petroleum and natural gas sector.

Petroleum and natural gas systems are heavily dependent on commercial electricity during the entire cycle of production, refining, processing, transport, and delivery to the ultimate consumer. The availability of commercial power is the most important dependency for the domestic oil sector. The natural gas infrastructure depends on electric power to operate lube pumps for compressors, after-cooler fans, electronic control panels, voice and data telecommunication, computers, SCADA communication and controls infrastructure, gas control centers, and other critical components.

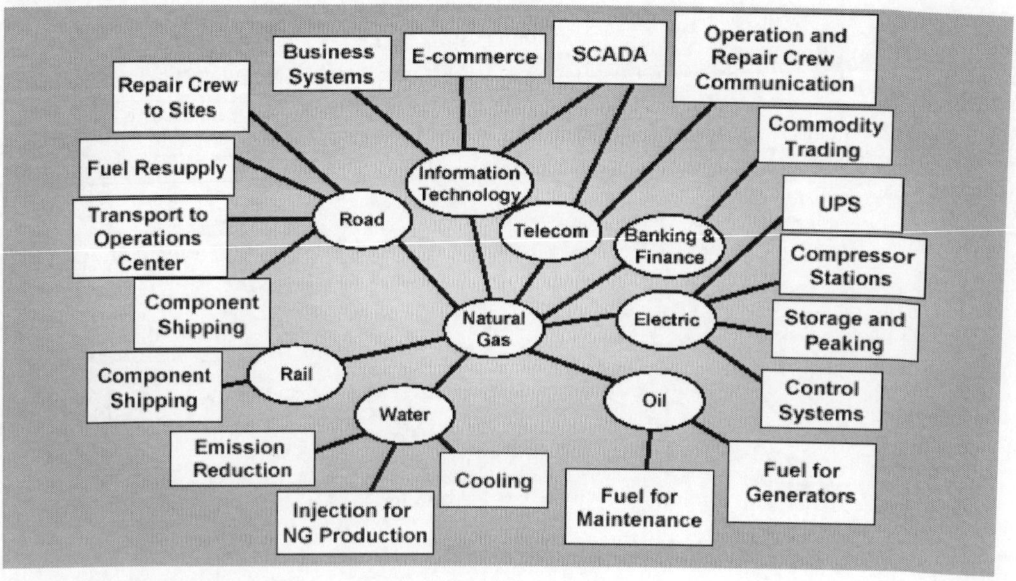

Figure 5-6. Examples of Natural Gas Interdependencies.

U.S. oil and natural gas companies operate a variety of telecommunications systems that are used to provide the internal communications capabilities that are crucial to protecting the safety of life, health, and property. These communications facilities are critical for the day-to-day operations of these companies, as well as for their response to potentially disastrous, life-threatening emergency situations. They are used for the direction of personnel and equipment, the control and synchronization of multiple geophysical acoustical signal sources for oil and gas exploration, and the telemetering of geophysical data. Mobile radio plays a critical role in providing communications for the management of individual wells; pipeline gathering systems; and in the transfer, loading, and delivery of petroleum products to end user consumers. In the event of emergency conditions, communication systems are essential to ensure the safety of personnel, the adjacent population, and the surrounding environment.

Petroleum and natural gas infrastructures are generally well equipped with gas-driven compressors and gas- or diesel-fired pumping facilities and backup generators that would enable the continued flow of natural gas, crude oil, and refined product deliveries for a limited time or that would implement a controlled shutdown following an interruption of electric power supply. There is also a possibility these backup generators may not function after an EMP event if they contain sensitive electronic components such as electronic control units. As one example of interdependency between the fuel and transportation sectors, we note that emergency generators that may keep critical electrical components of the petroleum and natural gas infrastructures running may become inoperative for lack of delivered fuel by a transportation sector short of fuels to run its trucks.

An electric power, water, or transportation disruption of short duration would not necessarily affect the operation of oil and natural gas infrastructure due to backup power and water resources. It is anticipated that crude oil and refined product deliveries could continue to flow for a few days, should these infrastructures be adversely affected. In the short term, natural gas deliveries are facilitated by the combined flexibility afforded by underground storage facilities and by line pack (the volume of gas maintained in the line at pressures above required delivery pressures). But outages of a few days or more can be expected to severely affect all infrastructure operations.

Recommendations

The Federal Government should take the lead in identifying this threat to the oil and gas industry sectors and specify ways to mitigate its potential consequences.

- The Energy Information Sharing and Analysis Center (ISAC) should, with government funding, expand its mission to address EMP issues relative to the petroleum and natural gas industries. This would include facilitating a government/industry partnership in addressing policy, investment prioritization, and science and technology issues.
- The Federal Government should review the feasibility of establishing a national inventory of component parts for those items that would be either in great demand or have long lead times, to be made available in a catastrophic event such as an EMP incident.
- Protect critical components.
  - The oil and natural gas industries should develop resource lists of existing SCADA and process control systems, with prearranged contracts and potential suppliers in the event of an EMP incident.
  - A study should be performed that prioritizes critical facilities of the oil and gas sector for future hardening against EMP effects.
  - Industry should strongly urge its members that have not already done so to install backup control centers to provide operational continuity. Industry should also explore the site location decisions for backup control centers so that adequate geographic separation between the main site and the backup facility is provided to protect against simultaneous damage in the event of a single EMP event.
- Develop training and exercises.
  - Individual companies should consider engaging in regional response and recovery planning and exercises to deal with disruptions to physical and cyber infrastructures resulting from an EMP event.
  - Emergency response manuals should be revised to include periodically recurring EMP event training for current and future work force.
  - Detailed simulation of the petroleum and natural gas infrastructure on a regional or local basis should be performed to provide a more accurate assessment of the potential impact of EMP-induced damage to these infrastructures.
- Conduct research.
  - Research and development efforts should stress hardening of SCADA and other digital control systems equipment, both existing and new components, to mitigate the impact of a future EMP event. New standards for oil and gas control systems should be established with the industry to avoid potential damage from EMP effects. These efforts could best be accomplished by the participation of

the various industry members, organizations (e.g., American Gas Association [AGA], Interstate Natural Gas Association of America [INGAA], Gas Technology Institute [GTI], American Petroleum Institute [API]), and government agencies.
- A cost-benefit analysis should be conducted for protecting the commercial petroleum and gas infrastructure against the effects of an EMP. If the costs are estimated to be substantial, the Federal Government should defray a portion of these costs.

SECTION 6. TRANSPORTATION INFRASTRUCTURE

Introduction

Transportation has played an essential role in our development from scattered settlements to a modern nation. Maritime (i.e., oceanic) shipping sustained the first settlements some five centuries ago and remains the most important avenue for intercontinental commerce today. The 18th century saw the rise of canals in the eastern states, and interest in them lasted through the first decades of the 19th century. Later the railroad supplanted canals in the east and opened the western territories for large-scale economic development and settlement. The 20th century witnessed the advent of the airplane and the automobile, both of which have radically transformed our economy and society. Water, rail, road, and air transportation now bind us together as a nation—economically, socially, and politically.

The criticality of transportation, the impact of potential disruptions, and the need to address vulnerabilities has received national attention. As recognized by the President's National Security Telecommunications Advisory Committee (NSTAC) Information Infrastructure Group Report:[68]

- The transportation industry is increasingly reliant on information technology (IT) and public information-transporting networks.
- Although a nationwide disruption of the transportation infrastructure may be unlikely, even a local or regional disruption could have a significant impact. Because of the diversity and redundancy of the United States (U.S.) transportation system, the infrastructure is not at risk of nationwide disruption resulting from information system failure. Nonetheless, a disruption of the transportation information infrastructure on a regional or local scale has potential for widespread economic and national security effects.
- Marketplace pressures and increasing use of IT make large-scale, multimodal disruptions more likely in the future. As the infrastructure becomes more interconnected and interdependent, the transportation industry will increasingly rely on IT to perform its most basic business functions. As this occurs, it becomes more likely that information system failures could result in large-scale disruptions of multiple modes of the transportation infrastructure.

- There is a need for a broad-based infrastructure assurance awareness program to assist all modes of transportation.
- The transportation industry could leverage ongoing research and development initiatives to improve the security of the transportation information infrastructure.
- There is a need for closer coordination between the transportation industry and other critical infrastructures.

The transportation sector of the economy is often addressed as a single infrastructure, but in reality its various modes provide for several separate, but related, infrastructures. Rail includes the long-haul railroad and commuter rail infrastructures, air includes the commercial and general aviation infrastructures, road includes the automobile and trucking infrastructures, and water includes both the maritime shipping and inland waterway infrastructures.[69] A combination of considerations—importance to the economy, potential for loss of life as a result of an electromagnetic pulse (EMP) attack, and criticality to civilian enterprises—has led us to focus on the long-haul railroad, trucking and automobile, maritime shipping, and commercial aviation infrastructures.

As far as transportation has developed, it is still far from static. The forces driving the continuing evolution of the transportation infrastructures can be understood in terms of the pursuit of competitive advantage, which derives from both lower cost and superior performance. Of particular importance, pressures for cost reduction have led to wide- spread adoption of just-in-time delivery practices. These practices not only reduce costs associated with maintaining large inventories, but also create strong dependencies on automated tracking of inventories and automatic sorting and loading to achieve efficient and reliable delivery of supplies and equipment. Just-in-time delivery is made possible by the application of technological advances in remote tracking, computer controls, data processing, inventory management, telecommunications, and uninterrupted movement. These technologies are all electronics-based and, hence, potentially vulnerable to EMP.

The imperative to achieve superior performance also has led to greater use of electronics, which has introduced a potential vulnerability to EMP. The automobile provides a familiar example of this phenomenon. Modern automobiles use electronics to increase engine performance, increase fuel efficiency, reduce emissions, increase diagnostic capability, and increase passenger safety and comfort.

The transportation infrastructures are trending toward increased use of electronics, thereby increasing potential EMP vulnerability.

To gauge the degree of vulnerability of the long-haul railway, trucking and automobile, maritime shipping, and commercial aviation infrastructures to EMP, the Commission has assessed selected components of these infrastructures that are vital to their operations. Our assessment is based on both data collected from testing conducted under the auspices of the Commission and other available test data that have direct applicability to transportation infrastructure assessment. For critical components of these infrastructures that we were unable to test—notably airplanes, air traffic control centers, locomotives, railroad control centers and signals, and ports—our assessment relies on surveys of equipment and communications links.

Long-Haul Railroad

Railroads excel at carrying voluminous or heavy freight over long distances. Class I railroad freight[70] in 2003 totaled some 1.8 billion tons originated.[71] The major categories of freight carried by railroads, illustrated in **figure 6-1**, include coal, chemicals, farm products, minerals, food products, and a variety of the other goods essential to the operation of our economy.[72]

Coal dominates all other categories of freight, accounting for 44 percent of Class I railroad tonnage in 2003. More than 90 percent of this coal, some 700 million tons, is delivered annually to coal-fired power plants. Power plants that depend on railroad-delivered coal account for more than one-third of our electricity production. Today, these plants typically have only several days' to a month's supply of coal on site. While this reserve provides a useful buffer, under conditions of a prolonged failure of railroads to deliver coal, these plants would simply have to shut down.[73] Electricity production would be affected most in the Midwest, Southeast, and Southwest, regions more heavily dependent on coal-fired power plants.[74]

Railroads have achieved significant gains in efficiency and safety by modernizing and automating their operations. Today, freight railways are controlled and operated from a limited number of centralized control centers. For example, the western U.S. Union Pacific tracks are managed from Omaha, NE, and the Burlington Northern/Santa Fe tracks are managed from Dallas, TX. These centers, as well as operations throughout the rail system, use extensive communication networks for sensing, monitoring, and control. If a railroad control center becomes inoperable or loses communications with the rail network for any reason, all rail traffic in the affected domain will stop until communications are restored or backup procedures are implemented.

EMP Vulnerability of the Long-Haul Railroad Infrastructure

The principal elements of the railroad infrastructure that we assessed are railroad control centers, railroad signal controls, and locomotives.

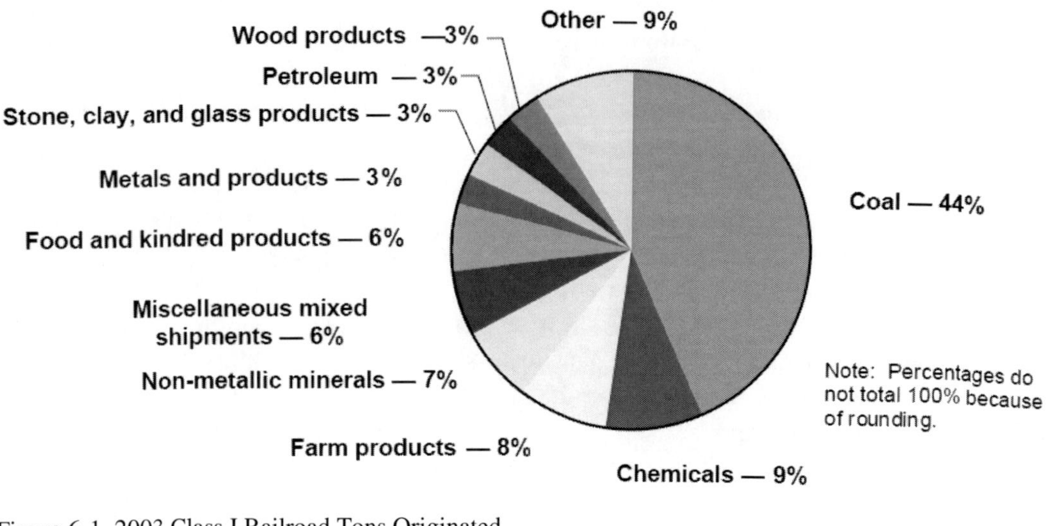

Figure 6-1. 2003 Class I Railroad Tons Originated.

Figure 6-2. CSXT Train Dispatch Center.

Railroad Control Centers

We conducted an EMP vulnerability survey of CSX Transportation (CSXT), the railroad subsidiary of the CSX Corporation. CSXT operates the largest rail network in the eastern United States. Like the other major railroad companies, CSXT has centralized its critical control facilities in a single geographical area. The CSXT Jacksonville, FL, rail- road control center includes three key nodes, each housed in a separate building—a customer service center, an advanced IT center, and a train dispatch center (**figure 6-2**). These buildings have no specific electromagnetic protection. About 1,200 trains are handled by the CSXT control center in a typical day.

Railroad control center operations rely on modern IT equipment—mainframe and personal computers, servers, routers, local area networks (LAN), tape storage units—some of which are similar to commercial off- the-shelf (COTS) equipment that has been EMP-tested. Based on this similarity, we expect anomalous responses of the IT equipment to begin at EMP field levels of approximately 4 to 8 kV/m. We expect damage to begin at fields of approximately 8 to 16 kV/m.

The CSXT railroad control center buildings rely on diesel power generators for standby power and central uninterruptible power supply (UPS) systems to provide continuous power to critical loads. Some buildings require chilled water for continuing computer operations. The buildings are interconnected by a fiber-optic ring and telephone lines. None of this equipment has specific EMP protection, and there are no data on the EMP vulnerability of this equipment.

The three railroad control center nodes are almost totally dependent on telephone lines (copper and fiber) for communications and data transfer. If all landlines fail, they still can communicate over a small number of satellite telephones, but data transfers would be severely limited.

Based on our assessment and test results, a weak link in the railroad infrastructure is the railroad signal controls, which can malfunction and slow railroad operations following exposure to EMP fields as low as a few kV/m.

Concerns about terrorist attacks and hurricanes have motivated CSXT to make provisions to operate for an extended period without support from the infrastructure. These provisions include diesel generators in case the two independent commercial power feeds should fail, fuel and food stored for 25 to 30 days of operation, beds for 50 people, and on-site wells to provide water.

In addition, all three of the key nodes have remote backup sites, either in Maryland or in the northern Midwest. This geographical dispersion provides some protection from a limited EMP attack. However, these backup sites rely on personnel in the Jacksonville area for operations at the remote sites, which makes them dependent on the infrastructure for transporting their personnel. It is possible that their personnel could be transported over the CSXT rail system if air and road transportation was interrupted by an EMP attack. They also are dependent on commercial telephone service to transfer the Jacksonville telephone numbers to the alternate sites and to establish the alternate data links.

In the case of EMP-caused outages of the three key facilities and the failure of the backup sites, railroad operations would be severely degraded. Customers could not place shipping orders, data processing would cease, and, most important, train orders could not be generated. Train orders define the makeup of trains, their routes, and their priorities on the track. Trains cannot operate without orders and would revert to fail-safe procedures. The first priority would be to stop the trains. If it were apparent that the outages would last for more than a few hours, efforts would be made to move the trains to the yards. This process could take up to 24 hours.

Once the trains and their crews are secured, plans would be made to resume operations under manual procedures. Implementation of manual procedures could take several days or longer, during which time it would be difficult to operate at more than approximately 10 to 20 percent of normal capacity. Train orders can be issued manually using satellite telephones. The biggest challenge is maintaining communications with trains that are underway. Train yards can communicate with trains by radio. If the trains are within about 20 miles of the yard, the entire communication path is wireless. However, longer- range communications use landlines to repeater stations along the train routes. The repeater station batteries provide only about 24 hours of standby power.

Shipment of critical supplies likely could resume under manual control operations. Transporting food from farms to storage warehouses and from storage warehouses to cities would be a high priority. Trains also deliver chemicals that cities use to purify drinking water and treat waste water. As discussed above, power plants generally have some reserve of coal on hand, but eventually it would become crucial to resume coal shipments to power plants.

Railroad Signal Controls

Railroads use two main types of controls: block controls and local controls. **Figure 6-3** shows a typical block signal control equipment enclosure and antenna. Block controls are used to assure that the next section (block) of track is clear before a train enters it. The main communications from the railroad control centers to the block controls uses a mix of radios and telephones. Block controls have battery backups that can sustain operations for up to 24 hours.

Local control systems manage grade crossings and signal both the train and the road traffic at a crossing. These control systems are designed to operate autonomously. Some modern local control systems have a minimal communications capability that consists of a telephone modem for fault reporting and possible downloading of programs and parameters for the controllers. Local control systems have battery back-up power, which would provide for normal operations from 8 to 48 hours, depending on the volume of train traffic.

Figure 6-4 shows a typical local grade crossing control shelter and sensor connection. Local control systems have sensors bolted or welded directly to the rails. The resistance of the circuit, closed by the train wheels and axle, is measured and used to predict the train's arrival at the crossing. Modern systems are in shielded steel enclosures that include extensive surge protection.

Similar electronics technologies are used in both road and rail signal controllers. Based on this similarity and previous test experience with these types of electronics, we expect malfunction of both block and local railroad signal controllers, with latching upset beginning at EMP field strengths of approximately 1 kV/m and permanent damage occurring in the 10 to 15 kV/m range.

The major effect of railroad signal controlfailures will be delayed traffic. For centrally controlled areas of track, if block signals were inoperative, manual block authority would be implemented. Where possible, signal teams would be sent out to manually control failed switches. Crews also would set up portable diesel units to power railroad crossings that had lost power. Railroad crossing generators are on hand for emergencies, such as hurricanes. Repair and recovery times will be on the order of days to weeks. If commercial power is unavailable for periods longer than approximately 24 hours, degraded railroad operations will persist under manual control until batteries or commercial power is restored.

Locomotives

We conducted an assessment of diesel-electric locomotives at the GE Transportation Systems plant (one of two manufacturers of diesel-electric locomotives) in Erie, PA. Our assessment is based on a review of locomotive construction practices, operational procedures, and limited test data. While we do not have direct test data on EMP effects on diesel-electric locomotives, some data are available from a test of a locomotive of different design that may provide some insight into the robustness of typical locomotive control electronics and subsystems.[75]

Figure 6-3. Typical Block Signal Control Equipment Enclosure.

Figure 6-4. Grade Crossing Shelter and Sensor Connection.

Two classes of locomotives were considered—those of the pre-microprocessor era and the more modern locomotives that make extensive use of electronic controls. Approximately 20 percent of the locomotive population is of the older generation; these are rapidly being replaced by the newer models. Electronics are not used to control critical functions in the older locomotives. We consider this generation of locomotives to be immune to EMP effects. While the locomotives themselves are considered immune, loss of communications with central dispatch or within the train requires that the engineer stop the train.

A block diagram showing the critical functions in the more modern locomotives is shown in **figure 6-5**. The major functions are traction (movement) and communications, both of which make extensive use of electronic components and, thus, are potentially vulnerable to EMP. As with older locomotives, the communications include communications to central dispatch and to other parts of the train. If these communications are lost for any reason, the train is required to stop.

The traction function is totally computer controlled, with the important exception of the engineer's emergency braking system. Three computers are used to control all major subsystems. Malfunction or loss of any of the computers will bring the train to a halt. Restoring operation could require the replacement of computers. Because few spare computers are provisioned, operations could be degraded until new computers are manufactured and installed—a process that could take months.

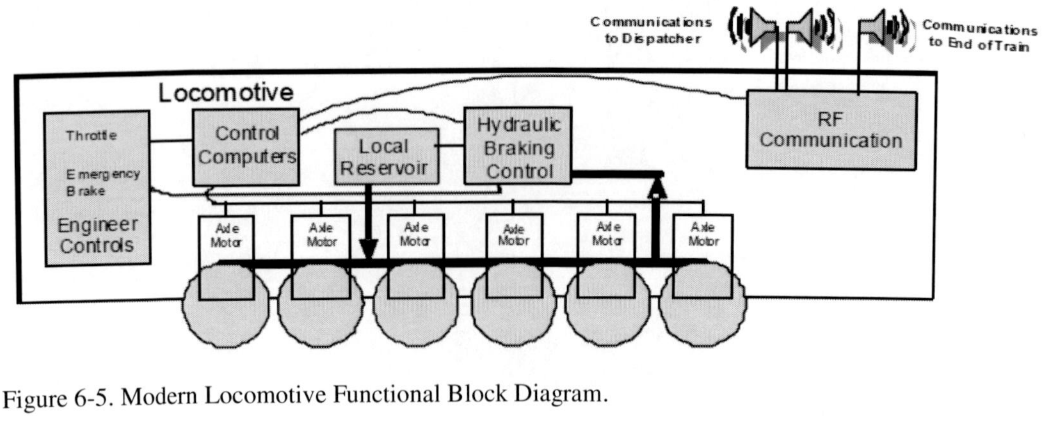

Figure 6-5. Modern Locomotive Functional Block Diagram.

It is important to note that computer failure or total loss of power in the locomotives could cause loss of electrical control for the brakes. In this case, there is a totally independent, nonelectrical system that the engineer can activate to apply the brakes in both the engine and the cars, thereby halting the train. Therefore, even in the worst case, the engineer can stop the train and prevent train crashes.

Because we did not directly test EMP effects on diesel-electric locomotives, the EMP vulnerability levels can be estimated based only on existing data for computer network response, locomotive construction methods, and the limited data available from the previously referenced test on an electro-mechanical locomotive belonging to the Swiss Federal Railroad.[76] Existing data for computer networks show that effects begin at field levels in the 4 to 8 kV/m range, and damage starts in the 8 to 16 kV/m range. For locomotive applications, the effects thresholds are expected to be somewhat higher because of the large metal locomotive mass and use of shielded cables. Therefore, we expect that effects will likely begin at incident field levels above the 20 to 40 kV/m range.

In summary, we consider the older generation of locomotives to be generally immune to EMP effects. Newer, electronically controlled locomotives are potentially more vulnerable. Based on construction practices, we expect that these vulnerabilities may manifest at EMP levels greater than 20 to 40 kV/m. While vulnerabilities may cause the locomotives to malfunction, fail-safe procedures ensure they can be stopped manually by engineers. Hence, we do not anticipate catastrophic loss of life following EMP exposure. Rather, we anticipate degraded operations, the severity of which depends on the incident EMP field levels. Normal locomotive operations can be restored on time scales from days to weeks or even longer. Restoration time scales could extend to months if computers, for which there are few spares, must be manufactured and replaced.

The Automobile and Trucking Infrastructures

Over the past century, our society and economy have developed in tandem with the automobile and trucking industries. As a consequence, we have become highly dependent on these infrastructures for maintaining our way of life.

Our land-use patterns, in particular, have been enabled by the automobile and trucking infrastructures. Distances between suburban housing developments, shopping centers, schools, and employment centers enforce a high dependence on the automobile. Suburbanites need their cars to get food from the grocery store, go to work, shop, obtain medical care, and myriad other activities of daily life. Rural Americans are just as dependent on automobiles, if not more so. Their needs are similar to those of suburbanites, and travel distances are greater. To the extent that city dwellers rely on available mass transit, they are less dependent on personal automobiles. But mass transit has been largely sup- planted by automobiles, except in a few of our largest cities.

As much as automobiles are important to maintaining our way of life, our very lives are dependent on the trucking industry. The heavy concentration of our population in urban and suburban areas has been enabled by the ability to continuously supply food from farms and processing centers far removed. Today, cities typically have a food supply of only several days available on grocery shelves for their customers. Replenishment of that food supply depends on a continuous flow of trucks from food processing centers to food distribution centers to warehouses and to grocery stores and restaurants. If urban food supply flow is substantially interrupted for an extended period of time, hunger and mass evacuation, even starvation and anarchy, could result.

Trucks also deliver other essentials. Fuel delivered to metropolitan areas through pipelines is not accessible to the public until it is distributed by tanker trucks to gas stations. Garbage removal, utility repair operations, fire equipment, and numerous other services are delivered using specially outfitted trucks. Nearly 80 percent of all manufactured goods at some point in the chain from manufacturer to consumer are transported by truck.

Our test results show that traffic light controllers will begin to malfunction following exposure to EMP fields as low as a few kV/m, thereby causing traffic congestion. Approximately 10 percent of the vehicles on the road will stop, at least temporarily, thereby possibly triggering accidents, as well as congestion, at field levels above 25 kV/m.

The consequences of an EMP attack on the automobile and trucking infrastructures would differ for the first day or so and in the longer term. An EMP attack will certainly immediately disable a portion of the 130 million cars and 90 million trucks in operation in the United States. Vehicles disabled while operating on the road can be expected to cause accidents. With modern traffic patterns, even a very small number of disabled vehicles or accidents can cause debilitating traffic jams. Moreover, failure of electronically based traffic control signals will exacerbate traffic congestion in metropolitan areas. In the aftermath of an EMP attack that occurs during working hours, with a large number of people taking to the road at the same time to try to get home, we can expect extreme traffic congestion. Eventually, however, people will get home and roads will be cleared as disabled cars are towed or pushed to the side of the road.

After the initial traffic congestion has subsided, the reconstitution of the automobile and trucking infrastructures will depend primarily on two factors—the availability of fuel and commercial power. Vehicles need fuel and service stations need electricity to power pumps. Few service stations have backup generators. Thus, replenishing the fuel supply and restoring commercial power will pace the return to normal operations. Similarly, restoration of traffic control systems will depend on the availability of commercial power and on the repair of damaged traffic control signals.

EMP Vulnerability of the Automobile and Trucking Infrastructures

We tested the EMP susceptibility of traffic light controllers, automobiles, and trucks.

Traffic Light Controllers

The road traffic control system is composed of sensors, control, and output systems. **Figure 6-6** shows a typical signalized intersection.

Control systems are implemented according to one of several specifications that have evolved over the years. We performed tests of 170E type controllers, in use in approximately 80 percent of signalized intersections. We tested a single controller box populated by multiple electronics cards. In the course of the testing, various cards were damaged and subsequently replaced to continue the testing. Four different types of effects were observed during intersection controller tests:

Figure 6-6. A Typical Signalized Intersection.

1. Forced Cycle: At field levels of 1 to 5 kV/m, the light was forced to cycle from green to red without going through yellow. This is a transient effect that recovers automatically after one cycle.
2. Disrupted Cycle: At field levels of 5 to 10 kV/m, the normally programmed cycle times became corrupted and change to a cycle different from that originally programmed. The controller had either been damaged or needed to be manually reset.
3. No Cycle: At 10 to 15 kV/m, the side street lights at an intersection never turned green. The controller had been damaged.
4. Flash Mode: Also at 10 to 15 kV/m, the intersection went into a mode in which the lights in all directions were flashing. This mode can cause large traffic jams because traffic flow is severely reduced in this situation. The controller has either been damaged or needs to be manually reset.

Based on these results, we anticipate that EMP will trigger moderate to severe traffic congestion in metropolitan areas. The traffic congestion may be exacerbated by the panic reactions possibly attendant to an EMP attack. None of the data predict or suggest life-threatening conditions; conflicting green lights did not occur during our tests. All the observed effects would cause less traffic disruption than would a power outage, which results in no working traffic lights.

The highway network's dependency on electrical power was demonstrated during Hurricane Isabel in 2003. Although some critical intersections were equipped with back-up power supplies, they typically were operational only for 24 hours. In many localities, during power outages, law enforcement officers were required to control the critical intersections. As such, these officers were taken away from other activities that they could be serving during emergencies.

Reestablishing normal traffic flow depends on the severity of the EMP-induced faults. Manual resets for all traffic signals in a medium-sized city (population of 500,000) can be accomplished in approximately a day, assuming available personnel.[77] The timeline for repairing damaged traffic controller boxes depends on the availability of spare parts. The timeline for either manual resets or repairs under stressed conditions are unknown.

Major metropolitan areas are establishing traffic operations centers (TOC) as an integral part of their traffic control infrastructure. A city's TOC is responsible for downloading the parameters controlling traffic signal timing and traffic signal coordination. However, a TOC is not a critical node from a traffic control standpoint. If the center were to become inoperable, the immediate effect would be on the city's integrated traffic system; the city would not be able to monitor its roadways, use its variable message signs along priority roadways such as interstates, or produce content for the cable channels or Internet updates that provide the public with information on traffic and highway conditions. The primary long term effect of a TOC failure would be a gradual drifting of the signal timing synchronization that the center provides to the intersections to which it connects.

Automobiles

The potential EMP vulnerability of automobiles derives from the use of built-in electronics that support multiple automotive functions. Electronic components were first introduced into automobiles in the late 1960s. As time passed and electronics technologies evolved, electronic applications in automobiles proliferated. Modern automobiles have as many as 100 microprocessors that control virtually all functions. While electronic applications have proliferated within automobiles, so too have application standards and electromagnetic interference and electromagnetic compatibility (EMI/EMC) practices. Thus, while it might be expected that increased EMP vulnerability would accompany the proliferated electronics applications, this trend, at least in part, is mitigated by the increased application of EMI/EMC practices.

We tested a sample of 37 cars in an EMP simulation laboratory, with automobile vintages ranging from 1986 through 2002. Automobiles of these vintages include extensive electronics and represent a significant fraction of automobiles on the road today. The testing was conducted by exposing running and nonrunning automobiles to sequentially increasing EMP field intensities. If anomalous response (either temporary or permanent) was observed, the testing of that particular automobile was stopped. If no anomalous response was observed, the testing was continued up to the field intensity limits of the simulation capability (approximately 50 kV/m).

Automobiles were subjected to EMP environments under both engine turned off and engine turned on conditions. No effects were subsequently observed in those automobiles that were not turned on during EMP exposure. The most serious effect observed on running automobiles was that the motors in three cars stopped at field strengths of approximately 30 kV/m or above. In an actual EMP exposure, these vehicles would glide to a stop and require the driver to restart them. Electronics in the dashboard of one automobile were damaged and required repair. Other effects were relatively minor. Twenty-five automobiles exhibited malfunctions that could be considered only a nuisance (e.g., blinking dashboard lights) and did not require driver intervention to correct. Eight of the 37 cars tested did not exhibit any anomalous response.

Based on these test results, we expect few automobile effects at EMP field levels below 25 kV/m. Approximately 10 percent or more of the automobiles exposed to higher field levels may experience serious EMP effects, including engine stall, that require driver intervention to correct. We further expect that at least two out of three automobiles on the road will manifest some nuisance response at these higher field levels. The serious malfunctions could trigger car crashes on U.S. highways; the nuisance malfunctions could exacerbate this condition. The ultimate result of automobile EMP exposure could be triggered crashes that damage many more vehicles than are damaged by the EMP, the consequent loss of life, and multiple injuries.

Trucks

As is the case for automobiles, the potential EMP vulnerability of trucks derives from the trend toward increasing use of electronics. We assessed the EMP vulnerability of trucks using an approach identical to that used for automobiles. Eighteen running and nonrunning trucks were exposed to simulated EMP in a laboratory. The intensity of the EMP fields was increase until either anomalous response was observed or simulator limits were reached. The trucks ranged from gasoline-powered pickup trucks to large diesel-powered tractors. Truck vintages ranged from 1991 to 2003.

Of the trucks that were not running during EMP exposure, none were subsequently affected during our test. Thirteen of the 18 trucks exhibited a response while running. Most seriously, three of the truck motors stopped. Two could be restarted immediately, but one required towing to a garage for repair. The other 10 trucks that responded exhibited relatively minor temporary responses that did not require driver intervention to correct. Five of the 18 trucks tested did not exhibit any anomalous response up to field strengths of approximately 50 kV/m.

Based on these test results, we expect few truck effects at EMP field levels below approximately 12 kV/m. At higher field levels, 70 percent or more of the trucks on the road will manifest some anomalous response following EMP exposure. Approximately 15 percent or more of the trucks will experience engine stall, sometimes with permanent damage that the driver cannot correct.

Similar to the case for automobiles, the EMP impact on trucks could trigger vehicle crashes on U.S. highways. As a result, many more vehicles could be damaged than those damaged directly by EMP exposure.

Maritime Shipping

The key elements of the maritime infrastructure are ocean-going ships and their ports. We did not perform an EMP assessment of ships.

There are more than 100 major public ports in the United States located along the Atlantic, Pacific, Gulf of Mexico, and Great Lakes coasts, as well as in Alaska, Hawaii, Puerto Rico, Guam, and the U.S. Virgin Islands. Deep-draft ports accommodate ocean- going vessels, which move more than 95 percent of U.S. overseas trade by weight and 75 percent by value.[78]

Ports handle a variety of cargo categorized as bulk cargo, including liquid bulk (e.g., petroleum) and dry bulk cargo (e.g., grain); break bulk cargo in barrels, pallets, and other packages; and general cargo in steel containers. Major commodities shipped through U.S. ports include:[79]

- Crude petroleum and petroleum products—oil and gasoline
- Chemicals and related products—fertilizer
- Coal—bituminous, metallurgical, and steam
- Food and farm products—wheat and wheat flour, corn, soybeans, rice, cotton, and coffee
- Forest products—lumber and wood chips
- Iron and steel
- Soil, sand, gravel, rock, and stone

Port Operations

Our assessment of maritime shipping infrastructure focuses on ports. EMP assessments were conducted for the Port of Baltimore in Maryland and ports in the Hampton Roads, VA, area. The Port of Baltimore assessment was performed at the Seagirt and Dundalk Marine Terminals. The assessment was hosted by the Maryland Port Administration. The Hampton Roads assessment was hosted by the U.S. Coast Guard (USCG) and conducted at their offices in Portsmouth, VA, and at the Norfolk International Terminal (NIT) in Norfolk—one of three terminals in the Hampton Roads area.

Under Coast Guard mandate, the National Vessel Movement Center (NVMC) was established to track notice of arrival information from ships entering all U.S. ports. The NVMC is located in Kearneyville, WV. All cargo ships greater than 300 gross tons must notify the NVMC at least 96 hours prior to their arrival.

For the ports of Baltimore and Hampton Roads, communications between ships and between ship and shore are primarily by way of very high frequency (VHF) radio. All vessels are required to monitor Channel 16 (156.8 MHz). A system of repeaters allows VHF communications 25 miles off shore. Some vessels have satellite communication systems. All vessels are brought into the ports by a pilot who boards the ship in open water.

Hampton Roads Area Port

NIT, one of the Hampton Roads area facilities, operates much like a bus stop. Ships with 2,000 to 4,000 containers arrive any hour of the day, any day of the week. A few hundred containers may be offloaded and additional containers loaded onboard. Then, after only 4 to 8 hours in port, the ship sails on to its next port. Most of the ships have regular routes. Some ships (15 percent) contain break bulk cargo, which is packaged cargo not in containers. The third type of cargo is bulk (like coal); however, NIT does not handle bulk cargo.

Containers are loaded on and off the vessels using sophisticated cranes designed specifically for the purpose (**figure 6-7**). The containers typically are loaded onto the chassis of yard trucks that shuttle them to storage locations around the port. In some cases "straddle carriers" are used instead of yard trucks.

Cranes are the key element in the operation of the terminal. The criticality of the cranes is underscored by the fact that repair crews are kept on site at NIT at all times. Repairs are required to be made in 15 minutes or less. Cranes have more than 100 computers and sensors in them. Replacement parts for normally anticipated failures are warehoused on site. However, the numbers of spares are not planned in anticipation of an EMP attack.

Each container has a unique identification number. The container number is noted when it is unloaded from a ship. When it is placed in the yard by one of the yard trucks (or straddle carriers), its parking place is sent to the data center in Portsmouth through a handheld wireless computer. All the container location data are mirrored to the data center at NIT and backed up daily. The data centers have UPS and diesel backup power. Personnel also walk the yard to reconfirm the accuracy and completeness of the container locations. There are typically 30,000 to 40,000 containers stored at NIT.

Eventually, the container is loaded onto a road truck or rail car for shipment to its destination. A container number is logged whenever the container passes through the entrance area. The final checkpoint has radiation detectors to look for radioactive materials that might be moved out of the terminal.

Port of Baltimore

The 275-acre Seagirt Marine Terminal is exclusively a container terminal. On the land side, containers arrive and leave primarily by truck (95 percent), even though the terminal is adjacent to CSX railroad's Intermodal Container Transfer Facility (ICTF). Seagirt has seven active electric cranes for loading and unloading ship containers. Like NIT, the Seagirt cranes rely on commercial power for their operation.

Nearby Dundalk Marine Terminal is more than twice as large (570 acres) and has a mixture of cargo types: passengers on cruise ships, containers, roll-on/roll-off (ro-ro), and break bulk. Dundalk does not process bulk cargo. The terminal has 10 dockside container cranes, which are of various vintages, all older than the Seagirt cranes. The Dundalk dockside cranes all use diesel-powered electric motors.

Dundalk docks on the channel next to Seagirt are used for ro-ro and break bulk cargos. Ro-ro cargos include automobiles and a large assortment of farm and construction equipment.

Both marine terminals use an assortment of diesel- and diesel/electric-powered equipment to move containers around the yard and onto and off of trucks and railroad cars. Diesel-powered top loaders are used to move and stack containers. **Figure 6-8** shows two of the six diesel/electric-powered rubber tire gantries (RTG) at Seagirt. They provide a more efficient method than the top loaders for moving and stacking containers. Unlike the dockside cranes, whose motion is limited by fixed rails, RTGs can be moved and placed strategically around the terminal.

Figure 6-7. Container Cranes and Stored Containers.

Figure 6-8. RTG at Seagirt Marine Terminal.

Information about the containers is transmitted to a central computer unit in the Seagirt computer room using wireless handheld Teklogix units (**figure 6-9**). Information about the status and storage location of each container is stored in the database using input from the handheld units. Conversely, the handheld unit operators can download information about any container from the central database. The container tracking systems at Seagirt and Dundalk are highly automated. Their operation is essentially paperless, which places heavy reliance on the integrity of the databases. To enhance reliability, all critical data are mirrored in near-real time to a nearby backup site (about 1 mile away). In addition, backup tapes are generated every evening. Seven days of backups are maintained at the backup site. The computer room uses a Liebert UPS for short-term backup power. Long- term emergency power is provided by a diesel generator. Because the current unit proved to be inadequate during a lightning-induced power outage, a new diesel generator is being installed. The new unit also will provide emergency power to critical equipment outside the computer room.

Figure 6-9. Handheld Wireless Data Unit.

Figure 6-10. Truck Control Station.

The land side of operations at Seagirt and Dundalk Marine Terminals is primarily concerned with controlling the ingress and egress of container trucks. Entry is regulated by a series of manned consoles overlooking the truck entry area (**figure 6-10**). Trucks pull up to speaker boxes where the driver provides information about the company, vehicle, and business at the terminal. Operators use remote cameras to read license plate numbers and other vehicle identification markings.

The operator enters the information into the database and is issued a routing slip that is printed near the speaker box. The slip looks similar to an airline boarding pass and contains information about the truck and the container with which it is concerned. The driver then proceeds to a manned checkpoint directly below the entry control consoles. Here, Seagirt personnel examine the routing slip and check the driver's identification before allowing the truck to proceed into the terminal to pick up or drop off a container. A similar check is performed when the truck leaves the terminal. All operations are entered into a database, providing real-time information on the status of each truck and its container. There are typically 1,600 truck operations a day at Seagirt.

EMP Vulnerability of Maritime Shipping

An EMP event could affect operations in every phase of the transfer of container cargo from ships at sea to the highways and rails of the United States. The ability to provide information on the cargo and crew 96 hours before reaching all ports in the United States could be degraded by EMP-induced failures at the NVMC. Even if the NVMC is not directly impacted by EMP, the ability of ships and their agents to communicate with the NVMC could be affected by a failure in the telephone system.

The USCG, under the authority of the captain of the port, can allow ships into port without a formal notification to the NVMC. The USCG would likely send one of its cutters to contact ships at sea by VHF radio. Its crew might board the ship and escort it to port. The choice of which ships to allow in and which to stop would be at the discretion of the USCG. Depending on the extent of the EMP-affected areas, ships might also be diverted to alternate ports.

Ships approved to enter port still need a pilot to navigate the inner waterways. Pilots use their own boats to reach the ships and use VHF radios for communication. It is unlikely that all the pilot boats and their radios would be damaged by an EMP event. There are always some that are not operating at any given time. Pilots normally rely on satellites for navigation, but they are capable of navigating using charts and buoys.[80]

An EMP event could slow down the arrival of ships to port, but it would not necessarily stop all arrivals. This was the case for the terminals in the Hampton Roads area after September 11, 2001. The terminals remained open, but the USCG was aggressive in boarding and escorting ships to port.

Dockside cranes are electrically powered from commercial power with no backup power source; loss of commercial power caused by EMP exposure would halt loading and unloading until electric power service is restored.

Once container ships are in port, they are dependent on the dockside cranes to load and unload containers. Most of the container cranes in the Hampton Roads area are powered by commercial power; the few remaining diesel-powered cranes are being replaced by electric cranes. All the dockside cranes at Seagirt also are powered by commercial power. The cranes using commercial power have no backup for commercial power. Thus, loading and unloading of containers would stop at these docks until commercial power is restored. The 10 dock- side cranes at Dundalk Marine Terminal are diesel/electric and independent of commercial power.

EMP might damage the container cranes. The cranes have myriad electrical components—programmable logic controllers, sensors, and motors. However, given their height, it is likely that they are struck frequently by lightning. While repair crews and replacement parts are kept on site at all times, these parts are unlikely to be sufficient to meet the replacement needs after an EMP attack.

Once containers are removed from a ship, they are placed in the yard in a numbered parking spot or in block storage, where canisters are stacked together like on a ship. Diesel powered yard trucks and straddle carriers are used for this purpose. It is unlikely that all of them would be damaged beyond repair by an EMP event. There are always units that are not operating, which, based on the test data taken on automobiles and trucks, would make them less likely to be damaged.

Equipment not damaged by EMP will be able to operate as long as it has diesel fuel. Typically, a 10-to-20 day supply of fuel is stored at the terminals. They normally rely on commercially powered electric pumps to move fuel out of the storage tanks, but would improvise alternate methods if there was an extended outage of commercial power.

The actual delivery and removal of containers from the ports is dependent on outside trucks and, to a lesser extent, railroads. Diesel/electric RTGs are used to move containers on and off trucks and rail cars. While RTGs are the most efficient method for moving containers, in the event they all failed, it is possible to load and unload containers with diesel-powered top loaders. Even ordinary forklifts could be used in an emergency. Ro-ro operations are less dependent on the operation of terminal equipment. Cranes are not used to unload the equipment—it just rolls down a ramp. Some break bulk cargo ships have their own cranes for dockside operations.

Container-handling equipment is only part of the port operations process. Record keeping is as important. Each container arriving at the port must be tracked until it leaves the port. If the records are lost, reconciling claims of lost containers could have a significant economic impact.

The location of each canister at the Hampton Roads area ports is stored in a database at a data center in Portsmouth. The data are mirrored to the data center at NIT and backed up daily. Both data centers have a UPS and backup generators. They rely on telephone lines to receive data and to communicate with each other.

It is unlikely that both data centers would be so damaged by EMP that they could not operate. They use multiple personal computers from different manufacturers to process the data. The NIT data center, which was visited as part of the assessment, uses Windows® software for some applications and Macintosh® software for others. This diversity in location, hardware, and software makes it less likely that there will be a total failure of the data processing system.

Even if all data on the container locations were to be lost, it would be possible to regenerate it in a few days. Personnel routinely roam the yard checking the accuracy of the database. They compare the container's unique number with the number of the parking spot. These personnel could reverse the process and regenerate the database.

The arrangement is similar at the Seagirt and Dundalk Marine Terminals. They have a central computer room with multiple servers that support the critical databases. The computer room also contains the base station for the wireless handheld units, various routers, and myriad telephone cables. There is no shielding that would limit EMP coupling to the large number of cables in the room. EMP-induced upsets should be expected and damage is certainly possible.

Critical data are mirrored to another data center about 1 mile away and backed up daily. Both data centers have a UPS and backup generators. The backup generator at Seagirt is inadequate to maintain operations and is being replaced with a more powerful unit that also will provide backup power to other critical equipment, such as the speakers and cameras at the truck gates.

It is unlikely that both of the Baltimore area data centers would be so damaged by EMP that they could not operate. They use multiple personal computers of different generations and from different manufacturers to process the data. The diversity in location and hardware makes it less likely that there will be a total failure of the data processing system. Paper records also would be needed to track the containers entering and leaving both the land and sea sides of the port. The ports could operate at significantly reduced capacity using a paper-based tracking system if necessary. It likely would take several days to implement the process.

Successful recovery from an EMP event will depend greatly on the availability of power and the ability of the USCG and port personnel to evaluate their situation and modify their operations accordingly. The events of September 11, 2001, and the need to survive periodic hurricanes have fostered the type of planning needed to respond to an EMP event. Although EMP was not directly considered, many of the plans for emergency recovery would be helpful after an EMP event.

During the assessment, it was encouraging that people in authority were clearly capable of responding well to unexpected situations. However, their response to an EMP event could improve significantly if they had a better understanding of what to expect.

Figure 6-11. An ARTCC Operations Room.

Commercial Aviation

Air travel has become ingrained in our way of life. There were 72 U.S.-certified airline carriers at the end of 2002, employing 642,000 pilots, flight attendants, mechanics, and other workers. U.S. airlines carried 560 million domestic passengers during 2001, logging some 700 billion passenger miles. In addition, U.S. airlines all carry freight to some extent. Commercial air freight shipments totaled about 22 billion ton-miles.[81]

The key elements of commercial aviation infrastructure that we assessed are the air traffic control system and the aircraft themselves.

The Federal Aviation Administration (FAA) has the responsibility for operating the U.S. air traffic control system with an emphasis on passenger safety. The FAA rigorously controls commercial air traffic—on the ground at airports (by airport towers), all takeoffs and landings (by Terminal Radar Approach CONtrol—TRACONs), and all en route travel (by air route traffic control centers—ARTCCs). Two essential parts of the FAA air traffic control architecture are (1) command and control through communication among controllers and between controllers and pilots, and (2) navigation aids for following proper routes, terminal approaches, and landing.

Commercial air traffic in U.S. airspace at altitudes up to 70,000 feet is controlled at all times. U.S. airspace is divided into 24 regions, 21 for the contiguous states and one each for Alaska, Hawaii, and Guam. Each region is controlled by an ARTCC. These centers provide en route control for aircraft at altitudes above 17,000 feet, maintaining safe separation between aircraft and routing aircraft around bad weather. Terminal control is provided for aircraft at lower altitudes and on the ground by airport towers.

An ARTCC has an operations room (**figure 6-11**) that consists of rows (banks) of individual controllers. The region controlled by a center is divided into sections. Aircraft are tracked and controlled by individual controllers and handed from controller to controller as the aircraft moves from section to section. Control passes from ARTCC to ARTCC over a dedicated private network telecommunications link that connects a controller from one facility to the next controller at another facility. En route control is acquired and handed off to a terminal controller in a similar manner.

If terminal control is interrupted, en route control takes over. If an en route control center is interrupted, control is turned over to another en route control center. These protocols provide redundant backup capability.

Radars are used to acquire and track aircraft in support of air traffic control centers. Generally, multiple radars will track an aircraft. Computers in air traffic control centers process radar information to form mosaic sectional displays and pass aircraft tracking information from center to center and across sections at a control center. Visualization is with a cathode ray tube (CRT) screen; paper printouts are also provided and used as a backup. Given a large number of radars with overlapping coverage, failure of a single radar will not adversely affect commercial air operations. Simultaneous failure of multiple radars, as could happen in an EMP attack, could shutdown all air traffic in the affected region, possibly nationwide.

The commercial aircraft in use are primarily jet-powered aircraft constructed by Boeing in the United States, and Airbus in Europe. In addition, there are various manufacturers of smaller commuter aircraft.

More than any other transportation infrastructure, the commercial aviation infrastructure is based on electronics. Everything from fly-by-wire aircraft flight control systems to navigation, communications, engine sensors and controls, and essential ground-based operations depends on microprocessor computer control.

Although a shutdown or curtailment of commercial aviation would have a severe, perhaps crippling, impact on the airline industry itself, the consequences for critical infrastructures would be less serious. Few vital economic activities are highly dependent on the unique advantage—speed—that commercial air transport has over the various modes of land transport.

EMP Vulnerability of the Commercial Aviation Infrastructure

Aircraft

Our commercial aircraft EMP assessment was conducted based on results of a meeting and subsequent discussions with Boeing electromagnetics effects (EME) staff. This staff is responsible for assuring that Boeing commercial aircraft can operate following exposure to nonhostile electromagnetic (EM) environments. Specifically, we assessed the amount of EMP protection that might be afforded by protection against lightning and high-intensity radiated fields (HIRF). Moreover, our assessment focused on safety of flight and the capability to land a plane after EMP exposure. We did not address continuation of normal flight operations, because we expect that all aircraft will be directed to land immediately on notification of an EMP attack.

Boeing maintains a strict engineering protocol for assuring their commercial aircraft are protected against nonhostile EM environments. This protocol includes qualification testing that is a function of flight-critical electronics categories, application of immunity standards to electronics boxes (sometimes referred to as line-replaceable units [LRU]), and hardening practices tailored to specific requirements.

EME Qualification Practices for Safety-of-Flight Electronics Boeing assigns electronics equipment to categories to differentiate the impact of loss of function. The highest category is reserved for electronics boxes, the failure of which would be considered catastrophic, and could lead to potential loss of the aircraft. Because our assessment focused on safety of flight, this is the most important category for EMP effects.

Although commercial aircraft have proven EM protection against naturally occurring EM environments, we cannot confirm safety of flight following EMP exposure. Moreover, if the complex air traffic control system is damaged by EMP, restoration of full services could take months or longer.

For this category of electronic subsystems, EME qualification is performed by a combination of low-level system tests and electronics box immunity tests (see next section). The purpose of the system-level tests is to estimate the intensity of the electromagnetic stresses coupled to the electronics box interfaces (connectors). For lightning (the EM environment most similar to EMP), the box immunity tests are then used to demonstrate that the electronics immunity levels are at least a factor of two higher than the coupled stresses. If this margin is not achieved, Boeing adjusts the protection tactics until this requirement is met. For lower-criticality electronic systems, only the box immunity tests are conducted, and there is no explicit relationship to the coupled stress required.

There has been significant evolution in the use of electronics in commercial aircraft. For aircraft designs prior to the 777, a direct mechanical/hydraulic link to the control surfaces was maintained, thereby minimizing electronics criticality for safety-of-flight applications. This observation would mitigate in favor of inherent EMP immunity for the nonelectronic subsystems. However, depending on aircraft, there are still some flight-critical functions performed by electronics, for which EMP immunity is not known. Therefore, even for pre-777 designs, there are insufficient data to confirm EMP immunity. Additional testing (limited to flight-critical electronics) is required to confirm EMP immunity. This testing should include low-level system testing to estimate EMP stresses at electronics interfaces and the corresponding electronics immunity testing. The recommended approach is essentially an extension of the existing lightning protocol to provide coverage for the EMP environment.

Boeing considers the 777 to be their first fly-by-wire design, incorporating more flight-critical electronics than used in earlier designs. Therefore, the newer designs may be more prone to EMP safety-of-flight impact. This potential is significantly mitigated by judicious use of redundancy for flight-critical subsystems. For example, while the flight-control systems use electrical signals rather than mechanical wires for control surface instructions, the primary digital controls are backed up by analog signals. Moreover, significant redundancy (up to four levels) is built into each flight-control subsystem. Therefore, the possible EMP susceptibility is offset significantly by careful, redundant design. Nonetheless, the qualification protocols do not provide adequate coverage for anticipated EMP responses. Therefore, as is the case for the earlier designs, additional testing is required to confirm EMP immunity. This testing should address both the EMP stresses at electronics interfaces and the corresponding immunity testing. Because there is more application of electronics in the newer designs, more extensive testing will be required than for the earlier designs.

EME Immunity Testing Standards The industry standard for electronics immunity testing for commercial aircraft is RTCA/DO-160D.[82] Boeing uses an internal standard that flows down from RTCA/DO-160D but is tailored to the company's technical practices. For lightning, damped sinusoid immunity testing at center frequencies of 1 and 10 MHz is required. Other EMP aircraft testing has shown that EMP response tends to be at higher frequencies, generally in the 10 to 100 MHz range. In addition, conducted susceptibility HIRF testing is required for frequencies covering and extending far beyond the EMP range. However, the test amplitudes are lower than might be expected for EMP. Therefore, EMP survivability cannot be directly inferred from commercial aircraft lightning and HIRF immunity testing standards.

EME Hardening Practices EME hardening in Boeing aircraft is achieved using a combination of tactics-stress reduction (e.g., use of shielded electrical cables), redundancy of flight-critical systems (depending on the system, up to four channels of redundancy are applied), and software error detection/correction algorithms for digital data processing. The combination of these tactics is adjusted to match the specific requirements of different electronic subsystems. In addition, hardening measures may be applied to electronic boxes to increase immunity, if required, to meet the Boeing specifications that flow down from DO-160D.

In summary, the Boeing engineering approach for protection and qualification against nonhostile electromagnetic environments is well established, and it is demonstrated by experience to be sufficient for the EM environments to which the aircraft are exposed during normal operations. While these procedures may provide significant protection in the event of an EMP attack, this position cannot be confirmed based on the existing qualification test protocols and immunity standards. This conclusion is applicable to all commercial aircraft currently in service, including the earlier designs. However, it is particularly emphasized for the newer, fly-by-wire designs that, by virtue of more reliance on digital electronics, may be more prone to EMP effects.

Air Traffic Control

We conducted an EMP vulnerability assessment of air traffic control by discussions with FAA engineers and former air traffic controllers and by visits to an FAA facility in Oklahoma City and the ARTCC in Longmont, CO. Moreover, because computer networks are integral parts of the air traffic control system, existing EMP test data on similar COTS electronics is applicable. Our testing did not include the FAA's private telecommunications network links connecting the ARTCCs, such as the FAA Leased Interfacility National Air Space Communications System (LINCS) and more recently the FAA Telecommunications Infrastructure (FTI) Program.[83] These FAA critical telecommunications networks and services are supported by a number of National Security and Emergency Preparedness (NS/EP) programs available from the Department of Homeland Security (DHS) National Communications System (NCS).[84]

The main function of ARTCCs is to control air traffic in surrounding regions. Regions are divided into sections and aircraft are monitored from section to section before being handed off to another ARTCC or to an airport approach control center. The process is highly computerized with quadruple computer redundancy and redundant power and internal communication systems.

The ARTCCs are composed, in part, of computer networks based on commercial components. Similar components have been EMP tested and have manifested latching upsets (requiring manual intervention to restore function) beginning in the 4 kV/m peak field range. Permanent damage has been observed in the 8 kV/m range but is more prevalent above 15 kV/m. Based on similarity, it is anticipated that ARTCCs will begin to manifest loss of function following EMP exposure to peak fields as low as 4 kV/m; but functions will not be seriously degraded unless exposed to peak fields in excess of 15 kV/m.

A large number of radars have overlapping coverage. Failure of a single radar will not significantly impact air traffic control capability. Simultaneous failure of multiple radars, as could happen in an EMP attack, could shutdown all air traffic control in the affected region, and possibly nationwide, thereby making it more difficult to assure safe landings. In this case, emphasis for safe landings would shift to aircraft crew and airport towers.

Power to all critical components of the FAA system is backed by fuel generator power, and in some instances, uninterrupted through temporary use of large UPSs. Visual flight operations will be in the forefront for collision avoidance and landing. Many aircraft will land at airports other than originally intended, as was the case after the 9/11 terrorist attacks. Significant challenges arise for safe landing in conditions of low visibility in the absence of navigation and landing aids at night and during adverse weather.

There are redundant radio communications with aircraft and redundant telephone and microwave communications between air traffic control regions and airport towers. If communications are lost, responsibility for safe landings will revert solely to the aircraft crews.

If the FAA air traffic control system is damaged by exposure to EMP environments, its reconstitution would take time. The FAA does not have sufficient staff or spare equipment to do a mass rapid repair of essential equipment. The FAA collection of radar, communication, navigation, and weather instruments spans 40 years. It includes components from multiple vendors that are connected using a variety of wire, wireless, and fiber links. Some equipment has lightning and electromagnetic interference protection. Accordingly, configuration control is difficult. It would take days to a month or more to bring various components of the control system back online, starting with communications, followed with navigation aids. As the control system rebuilds, there is likely to be significant reduction in air traffic, with constraints to increase intervals for departures, landings, and spacing of aircraft en route. Moreover, the capability to restore the air traffic control system is dependent on availability of services from other infrastructures. In the event these services are compromised by an EMP attack, the air traffic control restoration times will be extended.

Recommendations

Specific actions for each transportation infrastructure follow.

Railroads

Railroad operations are designed to continue under stressed conditions. Backup power and provisioning is provided for operations to continue for days or even weeks at reduced capacity. However, some existing emergency procedures, such as transferring operations to backup sites, rely on significant warning time, such as may be received in a weather forecast before a hurricane. An EMP attack may occur without warning, thereby compromising the viability of available emergency procedures. Our recommendations are directed toward mitigating this and other potential weaknesses. DHS should:

- Heighten railroad officials' awareness of the possibility of EMP attack, occurring without warning, that would produce wide-area, long-term disruption and damage to electronic systems.
- Perform a test-based EMP assessment of railroad traffic control centers. Develop and implement an EMP survivability plan that minimizes the potential for adverse long-term EMP effects. The emphasis of this effort should be on electronic control and telecommunication systems.
- Perform an EMP vulnerability assessment of current vintage railroad engines.
- Develop and implement an EMP survivability plan, if needed.

Trucking and Automobiles

Emphasizing prevention and emergency clearing of traffic congestion, DHS should coordinate a government and private sector program to:

- Initiate an outreach program to educate state and local authorities and traffic engineers on EMP effects and the expectation of traffic signal malfunctions, vehicle disruption and damage, and consequent traffic congestion.
- Work with municipalities to formulate recovery plans, including emergency clearing of traffic congestion and provisioning spare controller cards that could be used to repair controller boxes.
- Sponsor the development of economical protection modules—preliminary results for which are already available from Commission-sponsored research—that could be retrofitted into existing traffic signal controller boxes and installed in new controller boxes during manufacturing.

Maritime Shipping

The essential port operations to be safeguarded are ship traffic control, cargo loading and unloading, and cargo storage and movement (incoming and outgoing). Ship traffic control is provided by the Coast Guard, which has robust backup procedures in place. Cargo storage and movement is covered by other transportation infrastructure recommendations. Therefore, focusing on cargo operations in this area, DHS should coordinate a government and private sector program to:

- Heighten port officials' awareness of the wide geographic coverage of EMP fields, the risk caused by loss of commercial power for protracted time intervals, and the need to evaluate the practicality of providing emergency generators for at least some portion of port and cargo operations.
- Assess the vulnerability of electric-powered loading and unloading equipment.
- Review the electromagnetic protection already in place for lightning and require augmentation of this protection to provide significant EMP robustness.
- Coordinate findings with the real-time repair crews to ensure they are aware of the potential for EMP damage, and, based on the assessment results, recommend spares provisions so that repairs can be made in a timely manner.
- Assess port data centers for the potential of loss of data in electronic media.
- Provide useful measures of protection against EMP causing loss of function data.

- Provide protected off-line spare parts and computers sufficient for minimum essential operations.
- Provide survivable radio and satellite communication capabilities for the Coast Guard and the nation's ports.

Commercial Aviation

In priority order, commercial aviation must be assured that airplanes caught in the air during an EMP attack can land safely, that critical recovery assets are protected, and that contingency plans for an extended no-fly period are developed. Thus, DHS, working with the Department of Transportation, should:

- Coordinate a government program in cooperation with the FAA to perform an operational assessment of the air traffic control system to identify and provide the minimal essential capabilities necessary to return the air traffic control capability to at least a basic level of service after an EMP attack.
- Based on the results of this operational assessment, develop tactics for protection, operational workarounds, spares provisioning, and repairs to return to a minimum-essential service level.

All Transportation Sectors

- DHS should incorporate EMP effects assessment in existing risk assessment protocols.

SECTION 7. FOOD INFRASTRUCTURE

Introduction

A high-altitude electromagnetic pulse (EMP) attack can damage or disrupt the infrastructure that supplies food to the population of the United States. Food is vital for individual health and welfare and the functioning of the economy.

Dependence of Food on Other Infrastructures

The food infrastructure depends critically for its operation on electricity and on other infrastructures that rely on electricity. An EMP attack could disrupt, damage, or destroy these systems, which are necessary in making, processing, and distributing food.

Agriculture for growing all major crops requires large quantities of water, usually supplied through irrigation or other artificial means using electric pumps, valves, and other machinery to draw or redirect water from aquifers, aqueducts, and reservoirs. Tractors and farm equipment for plowing, planting, tending, and harvesting crops have electronic ignition systems and other electronic components. Farm machinery runs on gasoline and petroleum products supplied by pipelines, pumps, and transportation systems that run on electricity or that depend on electronic components. Fertilizers and insecticides that make possible high yields from croplands are manufactured and applied through means containing various electronic components. Egg farms and poultry farms typically sustain dense populations in carefully controlled environments using automated feeding, watering, and air conditioning systems. Dairy farms rely heavily on electrically powered equipment for milking cattle and for making other dairy products. These are just a few examples of how modern food production depends on electrical equipment and the electric power grid, which are both potentially vulnerable to EMP.

Food processing also requires electricity. Cleaning, sorting, packaging, and canning of all kinds of agricultural products are performed by electrically powered machinery. Butchering, cleaning, and packaging of poultry, pork, beef, fish, and other meat products also are typically automated operations, done on electrically driven processing lines. An EMP attack could render inoperable the electric equipment and automated systems that are ubiquitous and indispensable to the modern food processing industry.

Food distribution also depends heavily on electricity. Vast quantities of vegetables, fruits, and meats are stored in warehouses, where they are preserved by refrigeration systems, ready for distribution to supermarkets. Refrigerated trucks and trains are the main means of moving perishable foods to market; therefore, food distribution also has a critical dependence on the infrastructure for ground transportation. Ground transportation relies on the electric grid that powers electric trains; runs pipelines and pumping stations for gasoline; and powers signal lights, street lights, switching tracks, and other electronic equipment for regulating traffic on roads and rails.

Because supermarkets typically carry only enough food to supply local populations for 1 to 3 days and need to be resupplied continually from regional warehouses, transportation and distribution of food to supermarkets may be the weakest link in the food infrastructure in the event of an EMP attack. The trend toward modernization of supermarkets may exacerbate this problem by deliberately reducing the amount of food stored in supermarkets and regional warehouses in favor of a new just-in-time food distribution system. The new system relies on electronic databases to keep track of supermarket inventories so that they can be replaced with fresh foods exactly when needed, greatly reducing the need for large stocks of warehoused foods.

The electric power grid, on which the food infrastructure depends, has been component-tested and evaluated against EMP and is known to be vulnerable. Moreover, power grid blackouts induced by storms and mechanical failures on numerous occasions have caused massive failure of supermarket refrigeration systems and impeded transportation and distribution of food, resulting in spoilage of all perishable foods and causing food shortages lasting days or sometimes weeks. These storm- and accident-induced blackouts of the power grid are not likely to have consequences for the food infrastructure as severe or as geographically widespread as an EMP attack would.

In the face of some natural disasters like Hurricane Andrew in 1992, federal, state, and local emergency services combined have sometimes been hard pressed to provide the endangered population with food. Fortunately, there are few known instances of actual food starvation fatalities in the United States. In such localized emergencies as Hurricane Andrew, neighboring areas of the disaster area are usually able to provide needed emergency services (e.g., food, water, fire, and medical) in a timely fashion.

In the case of Hurricane Andrew, for example, although the area of the damage was relatively small, the level of damage was extraordinary and many people were affected. Consequently, emergency services were brought in, not just from neighboring states, but from many distant states. For example, electric transformers were brought in from other states to help rebuild the local power grid. The net result was a nationwide shortage of transformers for 1 year until replacements could be procured from overseas suppliers, who needed 6 months to build new transformers.

Hurricane Katrina, one of the greatest natural disasters ever to strike the United States, afflicted a much larger area than Andrew. Consequently, the ability to provide food and other emergency aid was a much greater challenge. The area disrupted by Hurricane Katrina is comparable to what can be expected from a small EMP attack.

Recent federal efforts to better protect the food infrastructure from terrorist attack tend to focus on preventing small-scale disruption of the food infrastructure, such as would result from terrorists poisoning some portion of the food supply. Yet an EMP attack potentially could disrupt or collapse the food infrastructure over a large region encompassing many cities for a protracted period of weeks, months, or even longer. Widespread damage of the infrastructures would impede the ability of undamaged fringe areas to aid in recovery. Therefore, it is highly possible that the recovery time would be very slow and the amount of human suffering great, including loss of life.

Making, Processing, and Distributing Food

The United States is a food superpower. It leads the world in production of the 10 major crops, nine of which are food sources: corn, soybeans, wheat, upland cotton, sorghum, barley, oats, rice, sunflowers, and peanuts. The United States is also a world leader in the production of meats, poultry, and fish. Of the world's 183 nations, only a few are net exporters of grain. The United States, Canada, Australia, and Argentina supply over 80 percent of the net cereal grains exported worldwide—the United States alone providing more than half.

These U.S. exports go far toward alleviating hunger and preserving political stability in many nations that lack the resources to feed their own populations. While most Americans tend to take for granted the quantity and high quality of food available to them on a daily basis, most other countries of the world regard the United States' food infrastructure as an enviable economic miracle.

In contrast to the United States, many of the world's nations struggle to meet the food demands of their populations, even though in some cases those populations are living near or below a subsistence level. Most of the world's 183 nations, to some degree, are dependent on food imports. Even among the advanced nations, the United States is exceptional for the quantity and quality of its food production.

U.S. consumers are supplied largely from domestically produced food. In 2002, according to data from the U.S. Department of Agriculture (USDA), some 2.1 million U.S. farms sold about $192 billion in crops and livestock. U.S. farms have 455 million acres under cultivation for crop production. Another 580 million acres in the United States are pasture and range land that support raising livestock.

Raw agricultural commodities are converted to intermediate foodstuffs and edible foods by some 29,000 processing plants located throughout the United States, according to the Census of Manufacturers. These plants employ about 1.7 million workers, which is approximately 10 percent of all U.S. manufacturing employment and just over 1 percent of all U.S. employment. Most plants are small, but larger establishments account for the major portion of shipments. The 20 largest firms in food manufacturing account for about 35 percent of shipments, while in beverage manufacturing, the 20 largest firms account for 66 percent of shipments. The largest 50 firms account for 51 percent of food shipments and 74 percent of beverage shipments.

Food is supplied to consumers by approximately 225,000 food stores, as well as by farmers markets and pick-your-own farms. Away-from-home food service is provided by approximately 850,000 establishments, including restaurants, cafeterias, fast food outlets, caterers, and others.

To illustrate how the U.S. food infrastructure works in making, processing, and distributing food from farm to market, here is a concrete example:

Washington State is the foremost apple producer in the United States, with more than $850 million in annual sales and 225,000 acres of orchards, mostly in the Cascade Mountains. A major supermarket chain contracts through a cooperative of medium-sized apple growers in the Spokane area to grow apples.

In the course of the growing season, the Spokane apple farmers use a wide array of farm machinery to tend their trees and to apply fertilizers and pesticides. During the harvest season, Washington farmers employ 35,000 to 45,000 pickers to harvest their apple crops. Hand-picked apples are loaded onto flatbed trucks and shipped to processing firms belonging to or under contract with the chain. Apples are processed on an electrically driven assembly line that uses a variety of electromechanical devices to clean fruit of dirt and pesticide residue, sort and grade apples according to size and quality, wax the fruit, and package it into 40-pound cartons.

If the apples are not to be sent to market immediately, they can be stored for up to 8 months in giant refrigerators. The chain arranges for a shipment of apples to its Maryland distribution center, located in Upper Marlboro, which services its stores in the Washing- ton, D.C., area. A trucking company is contracted for the 4-to-5 day shipment of apples to the East Coast using a refrigerated truck. The apples are offloaded at the Upper Marlboro regional distribution center, which makes daily deliveries to the chain's stores. A refrigerated truck delivers apples to a Washington, D.C., supermarket. Local residents purchase the apples.

This example of how the food infrastructure works for apples from grower to consumer generally is the same for most foods, with differences in detail. One important difference is that apples, compared to many other crops, are among those most dependent on manual labor and least dependent on machinery. Yet, clearly the food infrastructure, even for the apple, depends heavily on assembly lines, mechanical sorters and cleaners, refrigerators, and vehicles that, directly or indirectly, cannot operate without electricity.

Vulnerability to EMP

An EMP attack could damage or destroy some fraction of the myriad electronic systems, ubiquitous throughout the food infrastructure, that are essential to making, processing, and distributing food. Growing crops and raising livestock require vast quantities of water delivered by a water infrastructure that is largely electrically powered. Tractors, planters, harvesters, and other farm equipment are fueled by petroleum products supplied by pipelines, pumps, and transportation systems that run on electricity. Fertilizers, insecticides, and feeds that make possible high yields from crops and livestock are manufactured by plants requiring electric power.

Food processing—cleaning, sorting, packaging, and canning of all kinds of agricultural and meat products—is typically an automated operation, performed on assembly lines by electrically powered machinery.

Food distribution also depends on electricity. Refrigerated warehouses make possible the long-term storage of vast quantities of vegetables, fruits, and meats. Road and rail transportation depend on the electric grid that powers electric trains, runs pipelines and gas pumps, and powers the apparatus for regulating traffic on roads and rails.

Because the United States is a food superpower with relatively few farmers, technology is no longer merely a convenience—it is indispensable to the farmers who must feed the nation's population and much of the rest of the world.

In 1900, 39 percent of the U.S. population (about 30 million people) lived on farms; today that percentage has plummeted to less than 2 percent (only about 4.5 million people). The United States no longer has a large labor force skilled in farming that could be mobilized in an emergency. The transformation of the United States from a nation of farmers to a nation in which less than 2 percent of the population is able to feed the other 98 percent is made possible only by technology. Crippling that technology would be injurious to the food infrastructure with its security depending on the characteristics of an EMP attack.

The dependency of the U.S. food infrastructure on technology is much greater than implied by the reduction in the percentage of farmers from 39 percent in 1900 to less than 2 percent of the population today. Since 1900, the number of acres under cultivation in the United States has increased by only 6 percent, yet the U.S. population has grown from about 76 million people in 1900 to 300 million today. In order for a considerably reduced number of U.S. farmers to feed a U.S. national population that has grown roughly fourfold from approximately the same acreage that was under cultivation in 1900, the productivity of the modern U.S. farmer has had to increase by more than 50-fold. Technology, in the form of machines, modern fertilizers and pesticides, and high-yield crops and feeds, is the key to this revolution in food production. An attack that neutralized farming technology would depress U.S. food production.

The food processing industry is an obvious technological chokepoint in the U.S. food infrastructure. Food processing of vegetables, fruits, and all kind of meats is a highly automated, assembly-line operation, largely driven by electric power. An EMP attack that damages this machinery or blacks out the power grid would stop food processing. The work force in the food processing industry is sized and trained to run a largely automated system. In the event of an attack that stops the machines from running, personnel would not be sufficiently numerous or knowledgeable to process food the old-fashioned way, by hand. Depending on climate, most foods that are not refrigerated would begin to spoil in a few hours or days.

Finally, the distribution system is probably the most vulnerable technological chokepoint in the U.S. food infrastructure. Supermarkets typically carry only enough food to provision the local population for 1 to 3 days. Supermarkets replenish their stocks virtually on a daily basis from regional warehouses, which usually carry enough food to supply a multicounty area for about 1 month.

Regional warehouses are probably the United States' best near-term defense against a food shortage because of the enormous quantities of foodstuffs stored there. For example, one typical warehouse in New York City daily receives deliveries of food from more than 20 tractor trailers and redistributes to market more than 480,000 pounds of food. The warehouse is larger than several football fields, occupying more than 100,000 square feet. Packaged, canned, and fresh foods are stored in palletized stacks 35 feet high. Enormous refrigerators preserve vegetables, fruits, and meats and the entire facility is temperature controlled.

However, regional warehouses potentially are vulnerable to an attack that collapses the power grid and causes refrigeration and temperature controls to fail. Moreover, the large quantities of food kept in regional warehouses will do little to alleviate a crisis if it cannot be distributed to the population promptly. Distribution depends largely on trucks and a functioning transportation system. Yet storm-induced blackouts have caused widespread failure of commercial refrigeration systems and massive food spoilage.

Trends in the grocery industry toward just-in-time distribution may reduce reliance on regional warehouses and increase the vulnerability of the food infrastructure to EMP attack. Just-in-time distribution, now being adopted by some supermarket chains in California, Pennsylvania, and New Hampshire, uses automated databases and computer systems to track supermarket inventories in real time and promptly replenish food inventories, as needed, from even larger, but fewer, regional warehouses and directly from food manufacturers.

The new system promises to supply customers with fresher foods and to greatly reduce industry's reliance on large inventories of stockpiled foods at regional warehouses. As just-in-time distribution becomes the industry norm, in the event of an EMP attack, heavier reliance on computers and databases may make it easier to disrupt the management of food distribution, while decreased reliance on regional warehouses could greatly reduce the amount of food available for distribution in an emergency.

Pulse-current injection and free-field illumination testing on a limited number of refrigerators and freezers indicate that some units will fail from low to moderate EMP levels. This testing indicates that substantial numbers of people would have to survive without benefit of refrigerated foods for an extended period, until repairs or replacement refrigerators and freezers could be obtained. Massive food spoilage at stores and regional warehouses is implied.

Consequences of Food Infrastructure Failure

An EMP attack that disrupts the food infrastructure could pose a threat to life, industrial activity, and social order. Absolute deprivation of food, on average, will greatly diminish a person's capacity for physical work within a few days. After 4 to 5 days without food, the average person will suffer from impaired judgment and have difficulty performing simple intellectual tasks. After 2 weeks without food, the average person will be virtually incapacitated. Death typically results after 1 or 2 months without food.

This timeline would not start until food stockpiles in stores and homes were depleted. Many people have several days to weeks of food stored in their homes. For example, in 1996 when a snowstorm in the Washington, D.C., area virtually paralyzed the food infrastructure for a week, the general population was forced to live off of private food larders and had sufficient stores to see them through the emergency. However, a significant number of people, those with little or no home food supply, would have to begin looking for food immediately.

Historically, even the United States' vast agricultural wealth has not always been enough to protect its people from the effects of nature and bad economic decisions. Millions of Americans knew hunger as a consequence of a drought that caused the dust bowl years (1935 to 1938) in the Western and Central Plains breadbasket, as well as by the Wall Street crash of 1929 and the Great Depression. Even today, according to the USDA, 33.6 million Americans, almost 12 percent of the national population, live in "food-insecure households." Food-insecure households, as defined by the USDA, are households that are uncertain of having or are unable to acquire enough food to meet the nutritional needs of all their members because they have insufficient money or other resources.

A natural disaster or deliberate attack that makes food less available, or more expensive, would place at least America's poor, 33.6 million people, at grave risk. They would have the least food stockpiled at home and be the first to need food supplies. A work force preoccupied with finding food would be unable to perform its normal jobs. Social order likely would decay if a food shortage were protracted. A government that cannot supply the population with enough food to preserve health and life could face anarchy.

In the event of a crisis, often merely in the event of bad weather, supermarket shelves are quickly stripped as some people begin to hoard food. Hoarding deprives government of the opportunity to ration local food supplies to ensure that all people are adequately fed in the event of a food shortage. The ability to promptly replenish supermarket food supplies becomes imperative in order to avoid mass hunger.

Blackouts of the electric grid caused by storms or accidents have destroyed food supplies. An EMP attack that damages the power grid and denies electricity to warehouses or that directly damages refrigeration and temperature control systems could destroy most of the 30-day regional perishable food supply. Blackouts also have disrupted transportation systems and impeded the replenishment of local food supplies.

Federal, state, and local government agencies combined sometimes have had difficulty compensating for food shortages caused by storm-induced blackouts. For example:

- Hurricane Katrina in August 2005 caused a protracted blackout in New Orleans and the coastal region, destroying the food supply. Flooding, downed trees, and washed-out bridges paralyzed transportation. But the Katrina blackout by itself was sufficient to stop transportation and prevent rapid replenishment and repair of the food infrastructure because gas stations could not operate without electric power. An EMP attack could also paralyze transportation of food by rendering gas pumps inoperable, causing vehicles to fail and blacking out traffic lights, resulting in massive traffic jams. Hurricane Katrina's destruction of the food supply was a major contributing factor to the necessity of mass evacuation of New Orleans and the coastal population. Because many evacuees never returned, the protracted disruption of the food infrastructure, which lasted weeks—and in some localities months—while the electric power grid was being restored, was a major factor contributing to permanently reducing the populations of New Orleans and coastal Louisiana.

Hurricane Katrina's effect on the food infrastructure is comparable to what can be expected from a small EMP attack.

- Hurricane Lili in October 2002 blacked out the power grid in coastal Louisiana, virtually collapsing the local food infrastructure. As a consequence of the blackout, food was unavailable to thousands through normal means. In south Louisiana, 30 supermarkets would not open because the blackout prevented their electric cash registers from operating. Those stores that did open were stripped of food within hours. In Abbeville, the parking lots of shopping centers became feeding stations run by churches and the state Office of Emergency Preparedness. Associated Grocers, which supplies food to supermarkets in Louisiana, Texas, and Mississippi, sent food in refrigerated trucks to the area from regional warehouses.

  The food emergency was reflected in a skyrocketing demand for dry ice to preserve food stuffs during the hot weather and to preserve refrigerated foods. Local supplies of dry ice were exhausted quickly—one store selling 20,000 pounds of dry ice to hundreds of customers in 2 hours—and had to be supplemented with supplies from the Red Cross.

  It is important to note that no one died from food or water deprivation during this emergency, and that the damaged area was small enough to be aided rapidly during recovery by undamaged fringe areas.
- Hurricane Floyd in September 1999 put more than 200 supermarkets out of operation in North Carolina. Protracted blackouts caused massive food spoilage despite emergency efforts taken before the storm to preserve perishable goods in freezers. Floyd blackouts also impeded replenishment of some supermarkets by inducing traffic signal failures that contributed to massive traffic jams.
- An ice storm blacked out the Washington, D.C., area in January 1999. Warm food, potentially a survival issue in the freezing winter conditions, was not available in most people's homes because electric ovens and microwaves no longer worked.

  In addition, most gas-powered ovens would not work because those built since the mid-1980s have electronic ignition and cannot be lit with a match. Some resorted to cooking on camp stoves. Preserving refrigerated foods was also a concern that Pepco, the regional power authority, helped address by giving away 120,000 pounds of dry ice, all that it had. Dry ice became a precious commodity.
- In January 1998, an ice storm caused a widespread blackout affecting parts of Ontario and Quebec in Canada, and Maine and upstate New York in the United States. The blackout threatened the food supply. According to press reports, "Food poisoning has become a real threat as embattled Montrealers, unable to get to stores, eat food that has been kept too long in refrigerators that no longer work."

  In upstate New York, the electric utility Niagara Mohawk announced that it was focusing restoration of electric power on more populated areas "so that supermarkets, gasoline stations, and hotels could reopen, and people in the more rural areas could find food and shelter." New York State Electric and Gas helped customers get to shelters and distributed 200,000 pounds of dry ice for preserving food.
- Hurricane Andrew in August 1992 laid waste to 165 square miles in South Florida and left 3.3 million homes and businesses without electricity. Andrew's aftermath posed an immediate threat to life in South Florida, in part because of damage to the food infrastructure. Most grocery stores had been destroyed.

> Massive traffic jams, caused in part by nonfunctioning signal and street lights, prevented the surviving supermarkets from being resupplied. "More than 5,000 traffic lights are on the blink," the press reported. "Traffic was snarled for miles. The simplest chore, indeed almost everything, seemed to take forever."
> To meet the crisis, tons of surplus food were distributed in the area. Nonetheless, two weeks after the hurricane, food was still not reaching many victims.
> Andrew's blackout of the power grid made the crisis over food, water, and shelter worse by severing communications between relief workers and victims. Without power, there was an almost complete collapse of communications—no telephones, radio, or television. Consequently, many people were unaware of relief efforts or of where to go for help. Had Hurricane Andrew damaged a larger area, it is likely that undamaged fringe areas would have been less capable of coming to the rescue, resulting in a significant loss of life.

Storm-induced blackouts provide some basis for extrapolating the greater destructive effects on food infrastructure likely from an EMP attack. An EMP attack is likely to damage electric power grids and other systems over a much wider geographic area than blackouts caused by storms; therefore, recovery from an EMP attack probably would take longer. An EMP attack also could directly damage some electronic systems, including refrigeration systems and vehicles, which normally would not be damaged by a blackout. Compared to blackouts, an EMP attack could inflict damage over a wider geographic area and damage a much wider array of equipment; consequently, recovery of the food infrastructure from EMP is likely to be much more complicated and more protracted.

Federal, state, and local agencies combined would find it difficult to cope immediately or even over a protracted period of days or weeks following an EMP attack that causes the food infrastructure to fail across a broad geographic area encompassing one or more states. Infrastructure failure at the level of food distribution because of disruption of the transportation system, as is likely during an EMP attack, could bring on food shortages affecting the general population in as little as 24 hours.

Massive traffic jams are most likely in large cities, the very areas where rapid replenishment of the food supply at hundreds of supermarkets will be needed most urgently. Significantly, recent famines in the developing world have occurred, despite massive relief efforts by the international community, in large part because food relief could not reach victim populations through their underdeveloped transportation infrastructure. An EMP attack could, in effect, temporarily create in the United States the technological conditions in the food and transportation infrastructures that have resulted in developing world famines.

Recommendations

Current planning, as reflected in the President's National Strategy for the Physical Protection of Critical Infrastructures and Key Assets, the Public Health, Security, and Bioterrorism Preparedness and Response Act of 2002 (the Bioterrorism Act), and Federal Emergency Management Agency (FEMA) planning documents, all appear to assume relatively small-scale threats to the food infrastructure. Most concern is focused on terrorists' poisoning or infecting a small portion of the food supply to cause mass panic and public fear about the safety of all food. The FEMA Federal Response Plan for a food shortage assumes a disaster effecting about 10,000 people: "On the fringes of the geographic areas affected will be schools and small institutions having large inventories estimated to be sufficient to feed up to 10,000 people for 3 days and supply their fluid needs for 1 day."[85] Yet an EMP attack could so damage the food infrastructure that millions of people would be at risk. Recommendations to address this risk include the following:

- Relevant federal agencies, including the Department of Homeland Security and the USDA, should supplement their plans to meet food emergencies by drawing on federal food stockpiles.
- Federal food stockpiles should be sized to meet a possible large-scale food shortage in the event of massive disruption of the national food infrastructure from an EMP attack or other causes.
- The Federal Government should examine useful lessons learned from reviewing earlier plans and programs, such as those during the early Cold War years, when the Federal Government planned and prepared for food shortages on a large scale.
- The Federal Government should plan to locate, preserve, deliver, distribute, and ration existing stockpiles of processed and unprocessed food, including food stockpiles by the USDA and other government agencies, which will be an important component of maintaining the food supply.
- The Federal Government should make it a priority to plan to protect, deliver, and ration food from regional warehouses, under conditions in which an EMP attack has disrupted the power, transportation, and other infrastructures for a protracted period.
- The Federal Government should make plans to process and deliver private and government grain stockpiles to significantly supplement the processed food stored in regional warehouses. According to the USDA's National Agricultural Statistical Service, total private grain stockpiles in the United States amount to more than 255 million metric tons. Federal grain stockpiles held by the Commodity Credit Corporation exceed 1.7 million metric tons, with 1.6 million metric tons of that amount dedicated to the Bill Emerson Humanitarian Trust for overseas emergency.
- The Federal Government should increase food stockpiles if existing stockpiles of food appear to be inadequate.
- Contingency plans also should be made to provide significant levels of personnel and technical support to speed the recovery of agriculture and food production from an EMP attack.

Presidential initiatives have designated the Department of Homeland Security as the lead agency responsible for the security of the food infrastructure, overseeing and working with the USDA. Currently, under the *Robert T. Stafford Disaster Relief and Emergency Assistance Act* (the *Stafford Act*) the President "is authorized and directed to assure that adequate stocks of food will be ready and conveniently available for emergency mass feeding or distribution" in the United States.[86] However, in practice, the *Stafford Act* has been used to authorize purchasing food from private sources and issuing food coupons to be used in supermarkets in order to meet food shortages.

In some particularly dire emergencies, as during Hurricane Katrina and Hurricane Andrew, when private sector food resources were destroyed or inadequate to meet the crisis, the Federal Government has resorted to federal surplus foods. Many Andrew victims were saved from hunger by Meals Ready to Eat (MRE). But the Federal Government was surprised by Andrew, and the resort to MREs and surplus food stockpiles was a poorly planned act of desperation that came late in the crisis. Recommendations to achieve this initiative include the following:

- The Federal Government should consider one readily available option, which is to grow the food stockpile to include the MREs.
- Plans should include timely distribution of mass quantities of food, which is likely to be crucial during a shortage.
- The *Stafford Act* should be amended to provide for plans to locate, protect, and distribute existing private and government stockpiles of food and to provide plans for distributing existing food stockpiles to the general population in the event of a national emergency.

SECTION 8. WATER INFRASTRUCTURE

Introduction

Water and its system of supply is a vital infrastructure. High-altitude electromagnetic pulse (EMP) can damage or disrupt the infrastructure that supplies water to the population, agriculture, and industry of the United States (U.S.).

The water infrastructure depends for its operation on electricity. To the extent possible, aqueducts, tunnels, pipelines, and other water delivery systems are designed to rely on gravity. However, since the invention and proliferation of the electric water pump early in the last century, urban growth, planning, and architecture have been liberated from dependence on gravity-fed water systems. By making water move uphill, the gravity pump has made possible the construction and growth of cities and towns in locations that, in previous centuries, would have been impossible. Skyscrapers and high-rise buildings, which would be impractical if dependent on a gravity-fed water system, have been made possible by the electric pump.

Electrically driven pumps, valves, filters, and a wide variety of other electrical machinery are indispensable for the purification of water for drinking and industrial purposes and for delivering water to consumers. An EMP attack could degrade or damage these systems, affecting the delivery of water to a very large geographic region.

Electrical machinery is also indispensable to the removal and treatment of wastewater. An EMP attack that degraded the processes for removing and treating wastewater could quickly cause public health problems over a wide area.

Supervisory and Control Data Acquisition Systems (SCADA) are critical to the running and management of the infrastructure for delivery of pure water for drinking, for industry, and for the removal and treatment of wastewater. SCADAs enable centralized control and diagnostics of system problems and failures and have made possible the regulation and repair of the water infrastructure with a small fraction of the work force required in earlier days. As discussed in greater detail in Section 1, an EMP attack could damage or destroy SCADAs, making it difficult to manage the water infrastructure and to identify and diagnose system problems and overwhelming the small work force with systemwide electrical failures.

The electric power grid provides the energy that runs the water infrastructure. An EMP attack that disrupts or collapses the power grid would disrupt or stop the operation of the SCADAs and electrical machinery in the water infrastructure. Some water systems have emergency power generators, which could provide continued — albeit greatly reduced — water supply and wastewater operations for a short time.

Little analysis has been conducted of the potential vulnerability of the water infrastructure to EMP attack. However, SCADAs supporting the water infrastructure are known not to have been hardened, or in most cases even tested, against the effects of an EMP attack.

The electric power grid, on which the water infrastructure is critically dependent, is known to be vulnerable to feasible levels of EMP. Moreover, blackouts of the power grid induced by storms and mechanical failures are known to have disrupted the water infrastructure on numerous occasions. These storm- and accident-induced blackouts of the power grid are not likely to be as severe or as geographically widespread in their consequences for the water infrastructure as would an EMP attack.

Federal, state, and local emergency services, faced with the failure of the water infrastructure in a single large city, would be hard pressed to provide the population with the minimum water requirements necessary to sustain life over a time frame longer than a few days. They could not provide, on an extended emergency basis, the water requirements and services, including waste removal, necessary to sustain normal habitation and industrial production in a single large city; however, an EMP attack could disrupt the water infrastructure over a large geographic area encompassing many cities for a protracted period of weeks or even months.

The Water Works

Water for consumption and sanitation is taken for granted by virtually everyone in the United States. Yet, the infrastructure for supplying pure water to the U.S. population and industry and for removing and treating wastewater, compared to other infrastructures, took longer to build and arguably is the most important of all infrastructures for the sustainment of human life.

One of the most important differences between developed and underdeveloped nations is the availability of pure water. An estimated 1.3 billion people in the developing world, nearly one-quarter of the global population, lack access to safe drinking water and even more, approximately 1.8 billion, lack water for sanitation. Consequently, diseases related to impure water flourish in many underdeveloped nations, taking a devastating toll on health and longevity. Economic development in many developing world nations is impeded by the absence of an adequate water supply to support industry. Indeed, in some countries, a major obstacle to development is simply the fact that the labor force has no alternative but to spend much of its time transporting water for drinking and other domestic uses from distant and often contaminated sources.

In contrast to the water scarcity that impedes development in much of the developing world, the United States enjoys a healthy and growing population and economic prosperity supported by the efficient distribution and utilization of its abundant water resources. Freshwater consumption for all purposes averages about 1,300 gallons per capita per day in the United States. Irrigation and cooling account for about 80 percent of all consumption, and, in the 17 western states, irrigation alone accounts for more than 80 percent of water consumption. On average, some 100 gallons per person per day (200 gallons per person per day in the southwest) are consumed for domestic purposes such as drinking, bathing, preparing food, washing clothes and dishes, flushing toilets, washing cars, and watering lawns and gardens.

Drinking and cooking account for only a small fraction of the water consumed; however, because in most cases a single water source must serve all purposes, all water consumed, regardless of the purpose, must meet the standards for drinking water purity, as prescribed by law.

Supporting this demand for enormous quantities of high-quality water is a vast infrastructure that includes more than 75,000 dams and reservoirs; thousands of miles of pipes, aqueducts, and water distribution and sewer lines; 168,000 drinking water treatment facilities; and 19,500 wastewater treatment facilities.

A fairly small number of large drinking water and wastewater utilities located primarily in urban areas (about 15 percent of the systems) provide water services to more than 75 percent of the U.S. population.

There is no single organization or system controlling the entire water infrastructure of the United States. Rather, more than 100,000 utilities and private owners manage the national water infrastructure. However, because water utilities provide similar services and must meet similar standards, they all operate in much the same way.

Water supplies require collection, treatment, storage, and distribution. Surface water such as reservoirs, lakes, and rivers generally provides water for cities. Wells tapping underground aquifers often supply rural areas and the southwestern states. Homeowners with private wells typically drink the water directly, because the subsurface water has been filtered over many years within the natural underground sedimentation. Water treatment plants are designed to provide an uninterrupted water supply that raises the purity of surface water and aquifers to drinking standards. A typical municipal water treatment plant purifies water through several steps: filtration, coagulation, flocculation, sedimentation, and disinfection.

Filtration by utilities passes raw water first through coarse filters to remove sticks, leaves, and other large debris. Finer filtration passes water through layers of sand and other granular material to remove silt and microorganisms. This stage of treatment imitates the natural filtration of water as it moves through the ground. This entire process is accomplished through low-lift pumps and mechanically cleaned bar screens and fine screens.

Coagulation is the process of removing colloidal impurities, finely suspended particles, from the filtered water. A coagulant, such as aluminum sulfate, is thoroughly mixed into water containing colloidal particles. Aluminum sulfate not only will coagulate and remove colloidal particles, but also will react with calcium hydroxide in the water, forming aluminum hydroxide, which can be removed through further filtration or sedimentation.

Flocculation immediately follows the coagulation process to remove the finest particles that would never settle out naturally. The velocity of the water is reduced and a gentle mixing action is used to allow formation of insoluble salts, colloidal particles, and other remaining suspended matter into a “floc” particle. The colloids and the coagulants mix with each other to form a large neutral floc particle that will settle out during sedimentation.

Sedimentation involves moving the water to large tanks to allow the floc to settle to the bottom of the tank. Sedimentation basins or clarifiers are usually the largest tanks in the treatment process. About 1 pound of sludge is created for every pound of chemical added to the water for coagulation and flocculation. The sludge must be removed and disposed of and filters and screens must be backwashed regularly.

Disinfection uses chemicals to kill any microorganisms that may have survived the filtration process. Chlorine is the most common disinfectant. When chlorine combines with organic material, such as dead leaves, it produces potentially dangerous trihalomethanes (THM). Large water treatment plants in major cities often undertake an additional purification step that reduces the level of THMs. Ozone oxidation and ultraviolet light are other disinfectant processes that are sometimes used instead of or in addition to chlorine. Fluoride also may be added because of its ability to retard tooth decay. Groundwater is often aerated by bubbling air through the water or by spraying to oxidize dissolved iron and manganese and to remove odors caused by hydrogen sulfide.

Treated water is delivered by high-lift pumps to the distribution system, usually through pipelines pressurized to 40 to 80 psi, to consumers. These pumps help to maintain water levels in storage reservoirs. Gravity flow, whenever possible, is the preferred method for delivering water. However, most water must be delivered by means of electric pumps. High-pressure pumps at the treatment plant deliver water to various zones within a water district to a booster pump or series of booster pumps that completes delivery to the consumer. High-rise buildings typically are serviced by individual booster pumps with enough pressure to provide water to rooftop reservoirs for consumption by upper floors and to provide water for firefighting.

Many of the same processes used in purification of drinking water also are used in treatment of wastewater, suitably modified for the removal of the greater amounts of material found in sewage. Sewage provides an ideal environment for a vast array of microbes, primarily bacteria, plus some viruses and protozoa. In fact, wastewater processing relies on benign microorganisms in the purification process. Sewage may also contain pathogens from the excreta of people with infectious diseases that can be transmitted by contaminated water. Waterborne diseases, while seldom a problem now in developed nations, are still a threat in developing countries where treated water is not available for public use.

Contaminants are generally removed from wastewater physically, biologically, and chemically. First, rags, sticks, and large solids are removed by coarse screens to protect the pumps. Then grit, the material that wears out equipment, is settled out in grit tanks or chambers. At this point, most of the small solids are still in suspension and can be removed and concentrated in the primary gravity settling tanks. The concentrated solids, called raw sludge, are pumped to an anaerobic digester for biological decomposition. The clarified effluent then flows to secondary treatment units for biological oxidation where the dissolved and colloidal matter in wastewater provides nutrients for microorganisms. A final gravity settling tank is used to remove microorganisms. This concentrated biological sludge is removed and returned to the anaerobic digester. Chemical disinfection, usually employing chlorine, is the last stage in the treatment of wastewater before it is discharged.

Vulnerability to EMP

The water infrastructure is a vast machine, powered partly by gravity but mostly by electricity. Electrically driven pumps, valves, filters, and a wide variety of other machinery and control mechanisms purify and deliver water to consumers and remove waste-water. An EMP attack could damage or destroy these systems, cutting off the water supply or poisoning the water supply with chemicals and pathogens from wastewater. For example:

- Total organic carbon (TOC) analyzers detect the levels of pollutants and pathogens in water. Determining water quality and the kind of purification treatment necessary depends on these sensors.
- Mechanical screens, filters, collector chains, skimmers, and backwash systems remove sludge and other solid wastes. Failure of these systems would pollute the water and quickly clog the pumps.
- SCADA systems enable remote control and instantaneous correction of potential problems with water quality, delivery, and wastewater removal and treatment. This process allows most water utilities to be nearly autonomous in operation, using a minimum number of personnel. In an emergency, such as an electrical blackout, some subsystems have been or could easily be modified for workarounds. For example, many valves have a manual bypass mode, and some water plants have emergency power generators. However, the efficiencies made possible by SCADAs have reduced the available number of trained personnel probably below the levels required for protracted manual operation of water treatment facilities. The failure of SCADAs would greatly impede all operations.
- High-lift and low-lift pumps are ubiquitous throughout the infrastructure for purifying and delivering water and removing wastewater. Water cannot be purified or delivered, nor sewage removed and treated, if these systems are damaged or destroyed.
- Paddle flocculators and other types of mixers are the primary means of chlorination and other chemical purification. If these systems cease functioning, water cannot be purified and likely would remain hazardous.

All of these systems depend on the electric grid for power. Large water treatment plants consume so much electricity, in some cases about 100 megawatts, that backup generators are impractical. For reliability, water treatment plants typically draw electricity from two local power plants. An EMP attack that collapses the electric power grid will also collapse the water infrastructure.

Consequences of Water Infrastructure Failure

By disrupting the water infrastructure, an EMP attack could pose a major threat to life, industrial activity, and social order. Denial of water can cause death in 3 to 4 days, depending on the climate and level of activity.

Stores typically stock enough consumable liquids to supply the normal demands of the local population for 1 to 3 days, although the demand for water and other consumable liquids would greatly increase if tap water were no longer available. Local water supplies would quickly disappear. Resupplying local stores with water would be difficult in the aftermath of an EMP attack that disrupts transportation systems, a likely condition if all critical infrastructures were disrupted.

People are likely to resort to drinking from lakes, streams, ponds, and other sources of surface water. Most surface water, especially in urban areas, is contaminated with wastes and pathogens and could cause serious illness if consumed. If water treatment and sewage plants cease operating, the concentration of wastes in surface water will certainly increase dramatically and make the risks of consuming surface water more hazardous.

One possible consequence of the failure of water treatment and sewage plants could be the release of sludge and other concentrated wastes and pathogens. Typical industrial wastes include cyanide, arsenic, mercury, cadmium, and other toxic chemicals.

Boiling water for purification would be difficult in the absence of electricity. Even most modern gas stoves require electricity for ignition and cannot be lighted by match. In any event, gas also may not be available to light the stoves (see Section 5). Boiling could be accomplished by open fires, fueled by wood or other flammables. Other possible mitigators are hand-held pump filters, water purification kits, iodine tablets, or a few drops of household bleach.

A prolonged water shortage may quickly lead to serious consequences. People preoccupied with finding or producing enough drinking water to sustain life would be unavailable to work at normal jobs. Most industrial processes require large quantities of water and would cease if the water infrastructure were to fail.

Demoralization and deterioration of social order can be expected to deepen if a water shortage is protracted. Anarchy will certainly loom if government cannot supply the population with enough water to preserve health and life.

The many homeowners with private wells also would face similar problems. There would be fewer workarounds to get their pumps operating again, if the pump controller is damaged or inoperable. Even if power is restored, it is unlikely the average homeowner would be technically competent to bypass a failed pump controller and figure out how to power the pump with bypass power lines.

The first priority would be meeting personal water needs. Federal, state, and local governments do not have the collective capability, if the water infrastructure fails over a large area, to supply enough water to the civilian population to preserve life.

Storm-induced blackouts of the electric grid have demonstrated that, in the absence of electric power, the water infrastructure will fail. Storm-induced blackouts have also demonstrated that, even in the face of merely local and small-scale failure of the water infrastructure, the combined efforts of government agencies at all levels are hard pressed to help. For example:

- Hurricane Katrina in August 2005 collapsed the water infrastructure in New Orleans and coastal Louisiana. The Katrina-induced blackout stopped the vast machinery for purifying and delivering water to the population. Water supplies were contaminated. The National Guard, among other resources, had to be mobilized to rush water and mobile water purification systems to the afflicted region. The water crisis—which was protracted because the blackout was protracted, the electric power grid requiring weeks and in some places months to repair—was a major contributing factor to the mass evacuation of the regional population. Once evacuated, many never returned. Thus the loss of water resources was a significant factor contributing to permanently reducing the population in the region. The effects of Hurricane Katrina on the water infrastructure are comparable to what can be expected from a small EMP attack.

- Hurricane Lili in October 2002 blacked out the power grid in coastal Louisiana. With no electricity, water pumps no longer worked, depriving the population of running water. Local bottled water supplies were quickly exhausted. Federal and state authorities resorted to using roadside parking lots and tanker trucks as water distribution centers.
- In September 1999, Hurricane Floyd blacked out electricity, causing water treatment and sewage plants to fail in some Virginia localities and, most notably, in Baltimore, Maryland. For several days, blackout-induced failure of Baltimore's Hampden sewage facility raised concerns about public health. With its three pumps inoperable, Hampden spilled 24 million gallons of waste into Baltimore's Jones Falls waterway and the Inner Harbor.
- An ice storm in January 1999 blacked out Canada's Ontario and Quebec provinces, causing an immediate and life-threatening emergency in Montreal's water supply, which depends on electricity for filtration and pumping. On January 9, the two water treatment plants that served 1.5 million people in the Montreal region failed, leaving the area with only enough water to last 4 to 8 hours. Government officials kept the water crisis secret, fearing public knowledge would exacerbate the crisis by water hoarding and panic. But as household water pipes went dry and reports of a water shortage spread, hoarding happened anyway and bottled water disappeared from stores. Warnings not to drink water without boiling it proved pointless, because people had no other way of getting water and no way to boil it in the mid-winter blackout. Montreal officials feared not only a shortage of drinking water, but also an inadequate supply of water for fighting fires. The Montreal fire department prepared to fight fires with a demolition crane instead of water, hoping that, if a building caught fire, the conflagration might be contained by demolishing surrounding structures. So desperate was the situation that provincial officials considered evacuating the city. Fortunately, Hydro-Quebec, the government's electric utility, managed to restore power to the filtration plants and restore water service before such extreme measures became necessary.
- In August 1996, a heat wave blacked out parts of the southwestern United States. Water supplies were interrupted in some regions because electric pumps would not work. Arizona, New Mexico, Oregon, Nevada, Texas, and Idaho experienced blackout-induced disruption in water service during the heat wave. In Fresno, where most of the city received water from wells powered by electric pumps, the city manager declared a local emergency. Only two of the city's 16 fire stations had water, and most of the fire hydrants were dry. Tankers were rushed in to supplement the fire department's water supply.
- Hurricane Andrew in August 1992 caused a blackout in South Florida that stopped water pumps from working. The blackout denied running water to hundreds of thousands of people stranded among the ruins left by Andrew, amidst Florida's

summer heat. To meet the immediate crisis more than 200,000 gallons of water were distributed. However, without electricity to power radio or television sets, mass communication virtually ceased to exist, and people were unaware of relief efforts or where to seek help. Thousands may have been saved from dehydration by pyramids of bottled water on street corners made free for the taking and by survivors who spread the word.

In all the examples cited, timely emergency services to provide water prevented loss of life from dehydration. However, had the outages lasted longer and the blacked-out areas been larger, the outcome could have been very different. Storms are merely suggestive of, and provide some basis for extrapolating, the greater destructive effects on water infrastructure likely from an EMP attack.

Storm-induced blackouts and their effects on the water infrastructure are an imperfect analogy to EMP attack. Taken at face value, storm-induced blackouts and their consequences for the water infrastructure grossly understate the threat posed by an EMP attack. Storms are much more limited in geographic scope compared to an EMP attack. Power grid and water infrastructure recovery from storms, compared to recovery from an EMP attack, is likely to happen more quickly because of the "edge effect"—the capability of neighboring localities and states to provide recovery assistance. Because an EMP attack is likely to damage or disrupt electronics over a much wider geographic area than storm-induced blackouts, rescuers from neighboring states and localities would face a much bigger job, and recovery of the water infrastructure would take a much longer time.

Nor do storm-induced blackouts replicate the damage from an EMP attack that may occur in small-scale electronic systems critical to the operation of the water infrastructure, such as electric pumps, SCADAs, and motor controls for filters and valves. Compared to storms, an EMP attack is likely to inflict not only more widespread damage geographically, but also deeper damage, affecting a much broader array of electronic equipment, which will contribute to a more complicated and protracted period of recovery.

Recommendations

A Presidential Directive establishes new national policy for protection of our nation's critical infrastructures against terrorist threats that could cause catastrophic health effects. National-level responsibilities have already been assigned to the Department of Homeland Security (DHS) and the Environmental Protection Agency (EPA) to protect the water infrastructure from terrorist threats. The EPA is the designated lead agency for protection of drinking water and water treatment systems. Under this directive:

- DHS and EPA should ensure that protection includes EMP attack among the recognized threats to the water infrastructure.
- The following initiatives should be amended:
  - The President's *National Strategy for the Physical Protection of Critical Infrastructures and Key Assets* (February 2003), which details a plan for protecting the United States' critical infrastructures, including the infrastructure for water. The President's plan:

    - Identifies threats to the water infrastructure as: "Physical damage or destruction of critical assets...actual or threatened contamination of the water supply...cyber attack on information management systems... interruption of services from another infrastructure."
    - Directs the EPA to work with the DHS, state and local governments, and the water sector industry to: "Identify high-priority vulnerabilities and improve site security...improve sector monitoring and analytic capabilities...improve sector- wide information exchange and coordinate contingency planning...work with other sectors to manage unique risks resulting from interdependencies."
    - Focuses on terrorism and threats other than EMP, but lends itself well (in particular, its structure and logic) to addressing any threat, and should be amended to include EMP.
  - The *Public Health and Bioterrorism Preparedness and Response Act of 2002 (Bioterrorism Act)*, signed into law by President Bush on June 12, 2002. The Bioterrorism Act:
    - Requires the authorities over many drinking water systems to conduct vulnerability assessments, certify and submit copies of their assessments to the EPA, and prepare or revise their emergency response plans.
    - Is concerned with terrorist contamination of drinking water with chemical or biological agents.
    - Could be amended to address the greater bio-chemical threat that an EMP attack potentially poses to the water supply than any of the threats envisioned in the *Bioterrorism Act* because an EMP attack that causes SCADAs in water treatment facilities to malfunction could release biochemical agents, and conceivably contaminate water supplies over a very wide region.
- DHS and EPA should follow the government-recommended emergency preparedness steps applicable to a wide range of civil emergencies arising from different threats. These steps include assuring availability of water during emergencies. To that end, the government has recommended that citizens stockpile both water supplies and means of purification. Implementing these recommendations will provide some measure of preparation for an EMP threat to the water supply.

SECTION 9. EMERGENCY SERVICES

Introduction

Emergency services are essential to the preservation of law and order, maintenance of public health and safety, and protection of property. Americans have come to rely on prompt and effective delivery of fire, police, rescue, and emergency medical services through local government systems. Backing up these local systems are state capabilities (e.g., state police and National Guard) and specialized capabilities such as those provided by the Department of

Homeland Security (DHS), the Department of Justice, the Centers for Disease Control and Prevention, and other federal entities.

Americans have come to rely on prompt and effective delivery of fire, police, rescue, and emergency medical services.

The demand for emergency services is large. Across the United States more than 200 million 9-1-1 calls are fielded annually.[87] Responding to these calls is an army of some 600,000 local law enforcement officers, 1 million firefighters, and more than 170,000 emergency medical technicians and paramedics.[88] Anticipated expenditures over the next 5 years for emergency response services are estimated at $26 billion to $76 billion at the state and local levels, supplemented by an additional $27 billion at the federal level.[89]

Emergency services at all levels are receiving increased emphasis as a consequence of the September 11, 2001, terrorist attacks. The focus is on preventing and responding to terrorism, including nuclear attack, but little emergency services planning specifically considers electromagnetic pulse (EMP) attack.

Little emergency services planning specifically considers EMP attack.

The primary focus in this Section is on local emergency services systems. In particular, this Section focuses on the communications systems to alert, dispatch, and monitor those emergency services. The great majority of resources are concentrated at the local level; state and federal assistance will likely be quite thin, given the large geographic extent of an EMP attack.

In addition to local emergency systems, we also address the federal Emergency Alert System (EAS), designed to serve the President and other leaders in communicating with the public in emergency situations. Although no President has ever used the EAS, it is reasonable to anticipate that it would be used in the event of an EMP attack.

Emergency Services Systems Architecture and Operations

Local Emergency Services Systems

Figure 9-1 depicts a generic modern local emergency service system. Shaded elements are those for which we have assessed EMP vulnerability, as discussed later in this Section.

Calls for assistance come in on cellular and land telephone lines to 9-1-1 operators at centers known as Public Safety Answering Points (PSAP). PSAPs typically include one or more 9-1-1 operators and dispatchers, communications equipment, computer terminals, and network servers. The 9-1-1 operator determines the service required and forwards the information for dispatch of the appropriate response units.

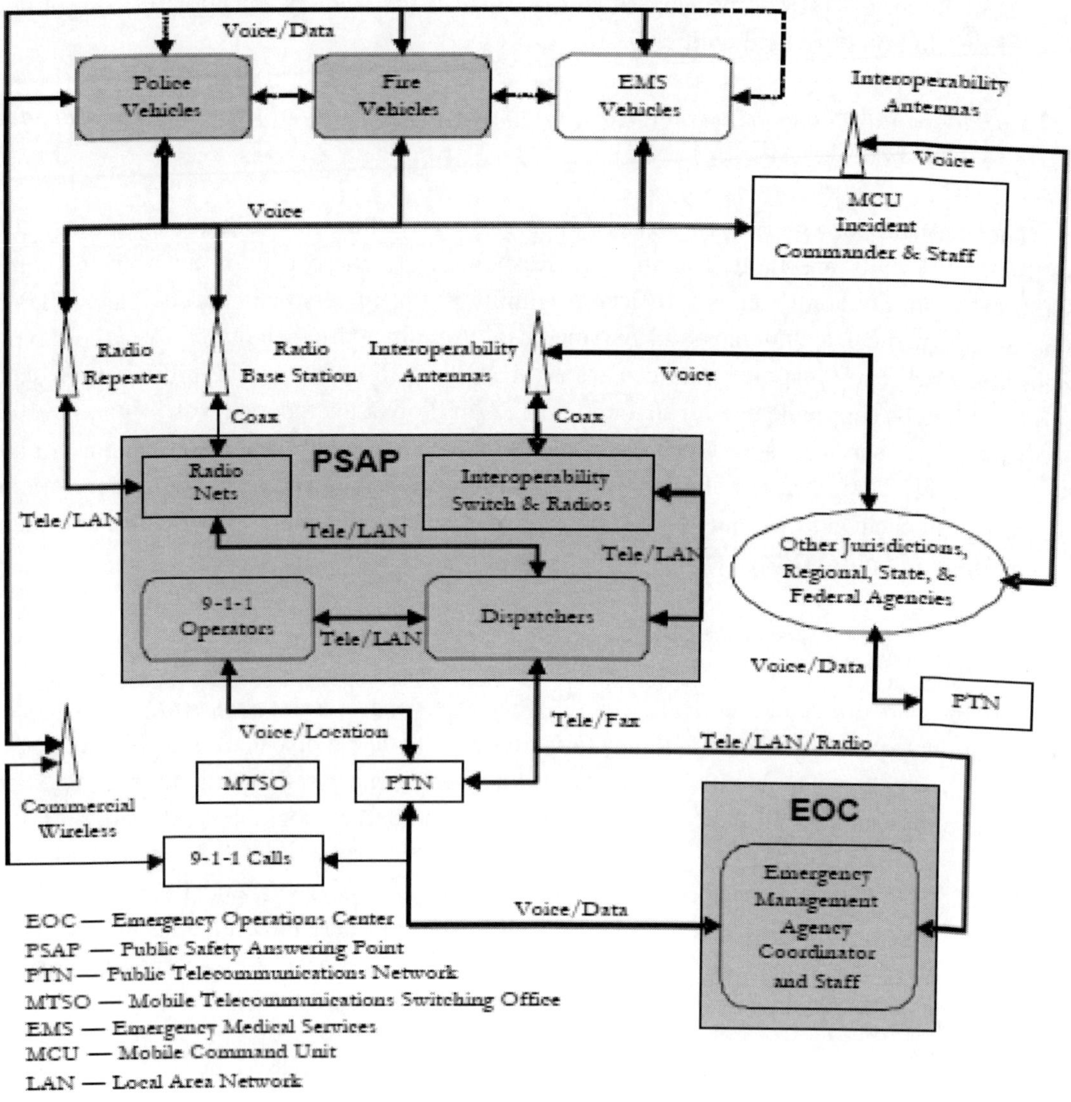

Figure 9-1. A Generic Modern Emergency Services System.

In addition to standard landline telephone service, emergency services employ a variety of wireless communication systems, including radio systems, cellular and satellite telephone systems, paging systems, messaging systems, and personal digital assistants. Because of dead zones and restrictions on radiated power levels in communications paths, radio repeaters are often used to relay voice and message traffic.

Because networks in nearby communities generally operate on different frequencies or channels to avoid interference, PSAP personnel use special equipment to handle community-to-community communications. If an emergency or public safety activity requires close and continuous coordination among several communities or agencies, an interoperability switch is used to allow direct communications among organizations. Interoperable communications across separate political jurisdictions is still a problem and under development in most regions.

For more serious emergencies, the Emergency Operations Center (EOC) serves as a central communications and coordination facility to which multiple organizations can send representatives. It facilitates efficient coordination across emergency services departments and state and federal agencies.

The Emergency Alert System

The original motivation for the EAS (previously the Emergency Broadcast System and initially Control of Electromagnetic Radiation [CONELRAD]) was to provide the President the ability to communicate directly with the American people in time of crisis, especially enemy attack. Although it has never been used for that purpose, it has been activated in local emergencies and is widely used for weather alerts. The Federal Communications Commission (FCC) sets requirements through regulation of television and radio stations. The Federal Emergency Management Agency (FEMA), now part of DHS, provides administrative oversight.

In the case of a national emergency, a message is relayed from the President or his agent to high-power amplitude modulation (AM) radio stations, known as national primary stations, across the country. These stations broadcast signals to other AM and frequency modulation (FM) radio stations, weather radio channels, and television stations that, in turn, relay the message to still other stations, including cable television stations. These stations use encoders and decoders to send and receive data recognized as emergency messages.

Impact of an EMP Attack

In a crisis, the priorities for emergency services are protection of lives, protection of property, effective communication with the public, maintenance of an operational EOC, effective communication among emergency workers, and rapid restoration of lost infrastructure capabilities. An EMP attack will adversely affect emergency services' ability to accomplish these objectives in two distinct ways: by increasing the demand for services and by decreasing the ability to deliver them.

The demand for emergency services will almost certainly increase dramatically in the aftermath of an EMP attack.

Demand for Emergency Services

The demand for emergency services will almost certainly increase dramatically in the aftermath of an EMP attack. These demands fall into two broad categories: *information* and *assistance.* The absence of timely information and the inability of recovery actions to meet the demand for emergency services will have grave consequences.

Large-scale natural and technological disasters that have occurred in the last several decades demonstrate that information demands are among the first priorities of disaster victims. At the onset of a disaster, an individual is concerned primarily with his or her personal well-being and that of close family members and friends. The next most pressing concern is for information regarding the event itself. What happened? How extensive is the damage? Who was responsible? Is the attack over? A less immediate priority is for information regarding recovery. How long will it take to restore essential services? What can and should I do for self-preservation and to contribute to recovery? It is important to recognize that emergency services providers also need all this information, for the same reasons as everyone else and also to manage recovery operations efficiently and perform their missions. Information assurance for emergency services requires reliable communications supporting the transport of emergency services such as enhanced 9-1-1 (E9-1-1).[90] As discussed in Section 3, Telecommunications:

> Based upon results of the Commission-sponsored analysis, an EMP attack would disrupt or damage a functionally significant fraction of the electronic circuits in the Nation's civilian telecommunications systems in the region exposed to EMP. The remaining operational networks would be subjected to high levels of call attempts for some period of time after the attack, leading to degraded telecommunications services.

To meet the demand for priority national security and emergency preparedness (NS/EP) services supporting first responders, the FCC and the DHS's National Communications System (NCS) offer a wide range of NS/EP communications services that support qualifying federal, state, and local government, industry, and nonprofit organization personnel in performing their NS/EP missions, including E9-1-1 PSAPs.[91,92]

The demand for assistance will increase greatly in the event of an EMP attack. The possibility of fires caused by electrical arcing resulting from an EMP attack cannot be ruled out. There is no reliable methodology to predict the frequency of such fires. As with other EMP effects, however, they will occur near-simultaneously, so that even a small number could overwhelm local fire departments' ability to respond. Fires indirectly caused by an EMP attack, principally because of people being careless with candles used for emergency lighting or with alternative heating sources during power blackouts, are also a concern.

There also exists the possibility of EMP-caused airplane crashes.[93] The average daily peak of air traffic in U.S. airspace includes more than 6,000 commercial aircraft carrying some 300,000 passengers and crew. Commercial aircraft are protected against lightning strikes but not specifically against EMP. The frequency composition of lightning and EMP differ enough so that lightning protection does not ensure EMP protection. On the other hand, the margins of safety for lightning protection imposed on commercial aircraft may provide flight safety in the event of an EMP attack. In any event, we cannot rule out the possibility of airplane crashes.

Debilitating EMP effects on the air traffic control system will also be a contributing factor to airplane crashes.

Emergency rescue services can be expected to experience an increase in demand. People trapped on subways and in elevators will require timely rescue. If electric power is interrupted for any period of time, people at home who depend on oxygen concentrators, respirators, aspirators, and other life-sustaining equipment that require electric power will need to find alternative solutions quickly. Home backup systems, including oxygen tanks, liquid oxygen supplies, and battery and generator power, will lessen the need for an immediate response for those fortunate enough to have them, but eventually all these people will need to be transported to facilities with a reliable power source and appropriate equipment. If power is out for more than several days, people dependent on dialysis machines, nebulizers, and other life-supporting medical devices also will be at risk. Finally, inability to replenish home supplies of medicines will eventually lead still more people to depend on emergency services.

Police services will be stretched extremely thin because of a combination of factors. Police will be called on to assist rescue workers in removing people from immediate dangers. Failures of automobiles and traffic control systems with attendant massive traffic jams will generate demands for police services for traffic management. Antisocial behavior also can occur following a chaotic event. Though it is more commonly seen in disasters originating from conflict, such as riots, than from natural or technological disasters, opportunistic crime (because of failures of electronic security devices, for example) is a potential reaction to an EMP attack. While not as prevalent as may be perceived, far worse antisocial behavior such as looting also could occur, especially in communities that experience conflict because of shortages or in areas that experience high crime rates under nondisaster circumstances. If looting or other forms of civil disorder break out, it is likely that local police services will be overwhelmed. In that event, deployment of National Guard forces, imposition of curfews, and other more drastic measures may be necessary.

Emergency services could be completely overwhelmed in the aftermath of an EMP attack.

Although emergency services could be completely overwhelmed in the aftermath of an EMP attack, it is important to recognize that the demand for emergency services could be ameliorated somewhat by citizen groups that frequently emerge in the aftermath of disasters to lead or assist in recovery efforts. In the absence or failure of government-provided emergency services, these groups may take on roles similar to those services, for example, by moving and providing basic household necessities to families in need, clearing debris, or serving as an impromptu communications network. This example of prosocial behavior is not uncommon in the aftermath of natural disasters such as hurricanes, floods, or earthquakes. This was seen following the September 11, 2001, terrorist attacks, when thousands of New York citizens volunteered to give blood, help firefighters and police at the World Trade Center grounds, and assist in other ways.

On the other hand, when the failure of police and emergency services becomes protracted, the lawless element of society may emerge. For example, Hurricane Katrina in August 2005 damaged cell phone towers and radio antennas that were crucial to the operation of emergency communications. Protracted blackout of the power grid caused generators supporting emergency communications to exhaust their fuel supplies or fail from overuse. Consequently, government, police, and emergency services were severely impacted in their ability to communicate with the public and with each other. Looting, violence, and other criminal activities were serious problems in the aftermath of Katrina. In one instance, the Danziger Bridge incident[94], members of a repair crew came under fire. Police called to the scene returned fire, and a number of people were killed. An EMP attack is likely to incapacitate the same nodes—cell phone towers and radio antennas— and overtax generators supporting emergency communications for a protracted period, creating the same conditions that incited lawless behavior in the aftermath of Katrina.

EMP Effects on Emergency Services

Some equipment needed to perform emergency services will be temporarily upset or directly damaged by an EMP attack, resulting in diminished capabilities during the time of greatest demand. Little, if any, emergency services equipment has been hardened specifically against EMP and thus may be vulnerable. On one hand, both communications equipment and vehicles commonly employed in the emergency services infrastructure generally have been designed to cope with the increasingly dense everyday electromagnetic environment from radio, television, wireless communications, radar, and other man-made sources. On the other hand, emergency services rely on radios to transmit and receive voice and message traffic using many frequencies, including the same frequencies contained in EMP radiation fields. Whether or not this results in degradation depends on the effectiveness of any built-in protection devices in these radios as well as the internal robustness of the radio itself.

An EMP attack will result in diminished capabilities during the time of greatest demand.

To gauge the degree of vulnerability of emergency services to EMP, the Commission conducted an assessment of emergency services equipment and associated networks.[95] We tested a representative variety of key electronics-based equipment needed by national leadership, first responders, and the general population. In most cases, only one of each model was tested, so statistical inferences are not possible from our test data. Moreover, a more robust assessment would test equipment under a range of conditions (such as different orientations, equipment operating modes, and test waveforms). Thus, our assessment should be viewed as indicative, rather than definitive. Notwithstanding these caveats, these tests are the most comprehensive recent vulnerability tests of emergency services equipment to date.

Our testing concentrated on items that were found to be critical for local emergency services and the EAS. The testing used standard EMP test practices, including radiated pulse and direct current injection test methods. Large-scale and smaller radiated pulse simulators were used to illuminate the equipment with an approximation of the electro-magnetic field generated by an actual EMP event. A second test method, known as pulse current injection, accounted for the stresses coupled to long lines such as power feeds that cannot be accurately tested in a radiated simulator. We also used the results of relevant past EMP testing efforts.

Public Safety Answering Points

The key elements of a PSAP include commercial telephone links for incoming 9-1-1 calls, computer-aided dispatch, public safety radio, and mobile data communications. There are other elements associated with PSAPs, but this is the minimal set necessary to provide emergency response to the public.

Computers are essential to normal PSAP operations. Recent personal computer equipment tests covered a wide technology range, consistent with what is typically in use in PSAPs. Results indicate that some computer failures can be expected at relatively low EMP field levels of 3 to 6 kilovolts per meter (kV/m). At higher field levels, additional failures are likely in computers, routers, network switches, and keyboards embedded in the computer-aided dispatch, public safety radio, and mobile data communications equipment.

A variety of mobile radios were tested in the stored, dormant, and operating states, in both handheld and vehicle-mounted configurations. Consistent with older test data,[96] none of the radios showed any damage with EMP fields up to 50 kV/m. While many of the operating radios experienced latching upsets at 50 kV/m field levels, these were correctable by turning power off and then on. However, most of the fixed installation public safety radio systems include telecommunication links between the computer-aided dispatch terminals and the main or repeater radio units. Therefore, because of computer failures in dispatch equipment, communication system failures might occur at EMP field levels as low as 3 to 6 kV/m.

Based on these results, we anticipate that several major functions of PSAPs will be affected by an EMP attack. The significance and duration of the impact of these failures will depend on multiple factors such as the ability of technical staff to repair or replace damaged equipment and the existence of plans and procedures to cope with the specific type of failure. For example, based on a review of representative Y2K public safety contingency plans, loss of the computer-aided dispatch capability can be overcome by the use of simple note cards for manually recording the information needed for dispatch. However, loss of the mobile radio communications or the incoming commercial telecommunications functions could be more difficult to counteract. Typically, local jurisdictions rely on nearby PSAPs or alternate locations to overcome these types of failures. In an EMP attack, these contingency plans may fail because of the wide area of effects.

Interoperability Switches

These switches are contained in many PSAPs to facilitate direct communications among local, regional, and state public safety departments and federal agencies after major disasters. The main elements of the interoperability switch capability are the public safety radios, the switch unit itself, and the computer network link between the switch unit and the dispatch console. The public safety radios that were tested as part of this assessment were based on the equipment used in a fully operational interoperability switch.[97] The testing was performed with the equipment in stored, dormant, and operating states. No failures were experienced at test levels up to 50 kV/m. The interoperability switch was also tested up to 50 kV/m with no adverse effects.

Based on these results, the interoperability switch capability is expected to function normally after an EMP attack. However, the computer network link between the interoperability switch and the dispatch station may fail at field levels as low at 3 to 6 kV/m. This would necessitate manual operation of the switch to implement the connections among various law enforcement, fire, and EMS agencies.

Vehicles

Emergency service vehicles include police cars, fire trucks, and EMS vehicles. An extensive test of a police car was performed. The most severe effect found was the latch-up of a mobile data computer at approximately 70 kV/m. After rebooting, the computer functioned normally.

Electronic equipment found on many of the mobile units also was tested. This equipment included a computer, personal data assistant, mobile and portable radios, defibrillators, and vital signs monitors. No permanent failures were experienced at levels up to 70 kV/m. Thus, we anticipate that the electronics in emergency services mobile units will continue to function normally, but they may suffer some initial effects due to latching upset of electronic devices.

Emergency Operation Centers

A site survey was performed at the Virginia state EOC. The survey confirmed that the vast majority of EOC communications depends on the Public Telecommunications Network (PTN). Thus, the ability of the EOC personnel to communicate and therefore provide emergency coordination will be highly dependent on the capability of the public telecommunications infrastructure to operate after an EMP event.

EOCs typically have at least one FEMA-owned and -maintained high-frequency (HF) radio for connectivity among national, regional, and state EOCs. The survivability of these HF radio units was not assessed. However, the operating band of these radios is one factor that makes them potentially vulnerable to EMP attack. Backup communications links may include satellite telephone systems and capabilities provided by amateur radio operator organizations.

EOCs also contain electronic equipment such as personal computers and digital data recorders. As with PSAPs, the capabilities supported by such equipment are vulnerable to EMP field levels as low as 3 to 6 kV/m.

Some EOCs are located below ground, which provides some protection from radiated EMP fields. However, conductive lines penetrating into these facilities must still be protected to ensure EMP survivability.

The Emergency Alert System

The primary method of initiating an emergency alert message involves the use of multiple commercial telecommunications lines. Therefore, the ability to provide emergency alert messages depends first on the status of the commercial telecommunications system. Broadcast of an alert message and receipt by the affected public depends on several electronic systems, including commercial radio and television stations, EAS multimodule receivers and encoders/decoders, and commercial radio and television receivers.

We performed site surveys of both a radio station and a television station. Backup power generators and spare transmitter equipment were found at both facilities. While not all commercial broadcast stations include such backup systems, the EAS has significant redundancy; some, but not all, broadcast stations are necessary for successful transmission of an emergency alert message.

We tested commonly used multimodule receiver and encoder/decoder units. The AM receiver module in its dormant mode failed at a field level of 44 kV/m. The FM receiver module exhibited erratic signal levels at 50 kV/m. No other effects were noted in testing EAS-specific equipment.

Four different television sets and two different radio receivers were tested. The vehicle testing performed for the transportation infrastructure assessment also tested radios in vehicles. In one AM radio installed in a vehicle, a malfunction occurred at approximately 40 kV/m. All other items showed no malfunctions.

Based on these results, we expect that the EAS will be able to function in near-normal fashion following an EMP attack. The major impact that might occur is a delay in initiation and receipt of an alert message because of (1) the dependency on the commercial telecommunications system, (2) the loss of some receiver channels for the EAS equipment, (3) the potential loss of some radio and television stations from power loss or damage to transmitter components, and (4) the loss of some AM radio receivers.

Interdependencies

In addition to direct damage, emergency services will be degraded to the extent that they are dependent on other infrastructures that are themselves damaged by the EMP attack. Emergency services are most directly dependent on the electric power, telecommunications, transportation, and fuel infrastructures. Fire departments also are dependent on the availability of water. EMP damage to these infrastructures can seriously degrade emergency services.

Of particular importance, emergency services are heavily dependent on the ability of the Nation's PTN to process 9-1-1 calls in a timely manner. After an EMP event, the PTN is likely to experience severe delays in processing calls.[98] Since 9-1-1 calls are processed using the same PTN equipment as non-9-1-1 calls (until they reach special 9-1-1 call- processing equipment located in a tandem central office assigned to each PSAP), they will be subject to delays similar to those for nonemergency calls. In the short term, this will result in a large number of lost 9-1-1 calls. After several days, the operation of the PTN is expected to return to near normal, assuming no adverse effects from either extended widespread power outages or from an inability to replenish fuel supplies for backup generators. However, in the event of a widespread power outage that extends beyond the time that backup power is available or commercial power service is restored, the PTN's ability to process 9-1-1 calls will again degrade. Eventually, extended wide- spread power outages will result in an inability to replenish fuel supplies, essentially causing a complete loss in PTN capability to process any 9-1-1 calls.

Loss of power can also directly impact PSAP operations. In the short term, the loss of commercial power will impact local emergency services more from the standpoint of increased calls for assistance than from functional impact. Most PSAPs and EOCs have backup power generators that will allow uninterrupted operation for some time period. Long-term power outages might result in the loss of PSAPs and EOCs because of an inability to refuel the backup generators.

Consequences

The ultimate consequences of an increased demand for emergency services and a concomitant degradation in emergency services capabilities are measured in lives lost, health impaired, and property damaged. We have no way of accurately estimating these consequences; we can only cite suggestive statistics.

We have no accurate way to measure the impact of degraded emergency services on lives lost, health impaired, or property damaged.

Most importantly, we note that the lives and health of many people depend on medical technologies that, in turn, depend on electric power. People will turn to emergency services if that power is unavailable for an extended period.

Emergency medical services respond to approximately 3 million 9-1-1 calls annually for people with cardiac problems and 2.5 million others for respiratory problems.[99]

Fire departments responded to 1,687,500 fires in 2002. These fires resulted in property damage estimated at $10.3 billion and 3,380 civilian deaths.[100] Lives and property saved by fire departments are undoubtedly also very large numbers.

Other direct consequences would result from the inability to successfully place a 9-1-1 call. Missed 9-1-1 calls can result from any number of causes, including (1) PTN outages; (2) EMP-induced damage to PSAPs, PSAP repeaters, mobile communications, or other critical support equipment; and (3) failure of commercial or residential telephone equipment.

The principal indirect consequences of a decline or collapse of emergency services are a result of a reduction in the availability of the work force. We did not attempt to quantify this effect, but note that it includes not only those directly affected, but also those who must now support those who previously would have depended on emergency services.

Recommendations

Our recommended strategy for protection and recovery of emergency services emphasizes the establishment of technical standards for EMP protection of critical equipment and the inclusion of EMP in planning and training.

The technology for critical emergency services functions is undergoing extensive change, creating an excellent opportunity for inclusion of our recommended protection measures. This technology change is propelled in large part by the need for additional emergency services communications capability and the recognition that large-scale dis- asters, such as the terrorist attacks of September 11, 2001, require extensive coordination across the full spectrum of emergency services providers.

Our recommended strategy for protection and recovery of emergency services emphasizes the establishment of technical standards for EMP protection of critical equipment and the inclusion of EMP in planning and training.

Our strategy can be realized through implementation of the following recommendations:

- DHS and state and local governments should augment existing plans and procedures to address both immediate and long-term emergency services response to EMP attack. Plans should include provisions for a protection and recovery protocol based on graceful degradation and rapid recovery that emphasizes a balance between limited hardening and provisioning of spare components. Such a plan should ensure the following:
  - The National Emergency Number Association should establish guidelines for operability and recovery of PSAPs during and after exposure to EMP.
  - The FCC should task the Network Reliability and Interoperability Council to address the NS/EP services, such as E9-1-1, and identify best practices to prevent, mitigate, and recover from an exposure to EMP.
- DHS should provide technical support, guidance, and assistance to state and local governments and federal departments and agencies to ensure the EMP survivability of critical emergency services networks and equipment. To accomplish this, the DHS should take the following actions:
  - In coordination with the Department of Energy and other relevant government entities, develop a set of EMP recovery scenarios that include coordinated attacks involving EMP and other more widely understood threats involving weapons of mass destruction.
  - In coordination with relevant government agencies, work with the appropriate standards entities (e.g., the Association of Public-Safety Communications Officials, the National Emergency Number Association, and the International

Electrotechnical Commission) to establish EMP immunity standards and guidelines for critical emergency services equipment.

- Develop training courses for emergency services providers on how to enhance immunity to, operate during, and recover from an EMP attack.
- Develop an EMP attack consequence assessment tool to perform planning analysis and training and to assist in the identification of critical equipment and manpower requirements.
- Establish a program to assess the vulnerability of evolving emergency services networks and electronics equipment to EMP and to develop a model plan for hardness maintenance and surveillance for implementation by state and local jurisdictions.

SECTION 10. SPACE SYSTEMS

Introduction

Over the past few years, there has been increased focus on U.S. space systems in low Earth orbits and their unique vulnerabilities, among which is their susceptibility to nuclear detonations at high altitudes—the same events that produce EMP. It is also important to include, for the protection of a satellite-based system in any orbit, its control system and ground infrastructure, including up-link and down-link facilities.

Commercial satellites support many significant services for the Federal Government, including communications, remote sensing, weather forecasting, and imaging. The national security and homeland security communities use commercial satellites for critical activities, including direct and backup communications, emergency response services, and continuity of operations during emergencies. Satellite services are important for national security and emergency preparedness telecommunications because of their ubiquity and separation from other communications infrastructures.

The Commission to Assess United States National Security Space Management and Organization conducted an assessment of space activities that support U.S. national security interests and concluded that space systems are vulnerable to a range of attacks due to their political and economic value.[101] Satellites in low Earth orbit generally are at risk of lifetime degradation or failure from collateral radiation effects arising from an EMP attack on ground targets.

In the course of an EMP attack, a nuclear detonation at a high altitude produces numerous other effects that can impact the performance and survival of satellites. Examination of these effects relates to the Commission's mandate in two ways. First, nuclear weapon effects on satellites can be collateral consequences of an EMP attack. Second, an EMP attack can degrade ground terminals that satellite systems require for uplinks, downlinks, and control functions.

This Section focuses on two classes of effects that are primary threats to the physical integrity of satellites: (1) direct, line-of-sight exposure to nuclear radiation pulses (e.g., X-ray, ultraviolet, gamma-ray, and neutron pulses) and (2) chronic exposure to enhanced high-energy electrons durably trapped in the Earth's magnetic field. These effects can jeopardize satellites in orbit, as data from U.S. and Soviet high-altitude nuclear tests of 1958 and 1962 attest. **Figure 10-1** illustrates visible phenomena from several U.S. high-altitude nuclear tests. Each detonation produced copious X-ray fluxes and trapped energetic electron radiation in space. When the United States detonated the 1.4-megaton (MT) STARFISH[102] device on July 9, 1962, at 400 km altitude, a total of 21 satellites were in orbit or were launched in weeks following. Eight suffered radiation damage that compromised or terminated their missions.[103] Information concerning the fate of the remaining 13 satellites is not publicly available.

In many respects, satellite electronics of the 1960s were relatively robust against nuclear effects. Their bulk and comparatively low-speed operation tended to make electronics of the era substantially less vulnerable to radiation upset and damage than modern electronics at comparable exposure levels. The discussion to follow highlights salient points of satellite vulnerabilities to nuclear explosions in the upper atmosphere or space. These vulnerabilities are considerable and incontrovertible — each worldwide fleet of satellites is at risk, but the degree of risk depends on the extent of satellite hardening, satellite location relative to the burst, resultant line-of-sight exposure to prompt radiations, and each satellite's exposure to geomagnetically trapped energetic particles of natural and nuclear origins.

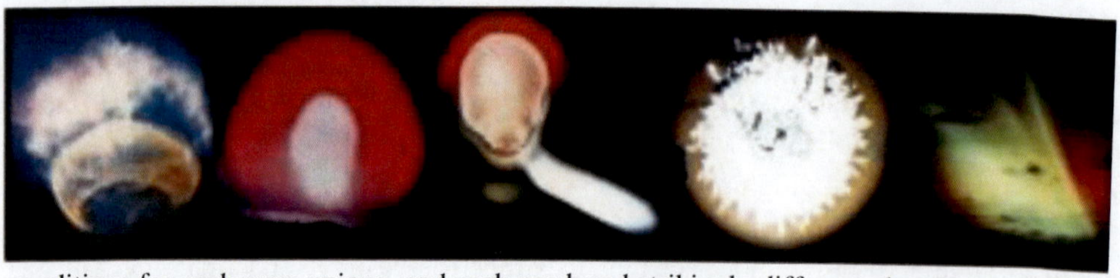

Burst conditions for each were unique, and each produced strikingly different phenomena and different enhancements of the radiation belts.

Figure 10-1. From left to right, the ORANGE, TEAK, KINGFISH, CHECKMATE, and STARFISH high- altitude nuclear tests conducted in 1958 and 1962 by the United States near Johnston Island in the mid-Pacific.

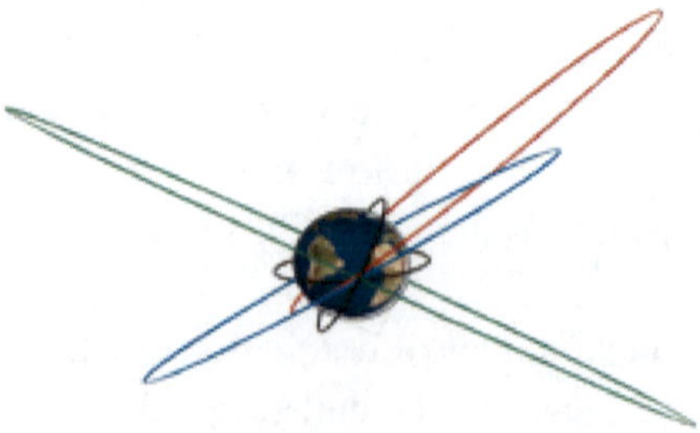

Geosynchronous orbit (green) in the equatorial plane is at about 36,000 km altitude. LEO (black) are shown with inclinations relative to the equatorial plane of 30° and 90°, but any inclination is possible. A 45° inclination orbit at approximately 20,000 km altitude is shown in blue. HEO are shown in red.

Figure 10-2. Satellite Orbits Illustrated.

Terms of Reference for Satellites

Ubiquitous Earth-orbiting satellites are a mainstay of modern critical national infrastructures. Satellites provide Earth observations, communications, navigation, weather information, and other capabilities. The United States experienced significant disruption when the pager functions of PanAmSat Galaxy IV failed in May 1998.

Each satellite's orbit is optimized for its intended mission. Low Earth orbits (LEO), from 200 to 2,000 km altitude, are in proximity to the Earth and atmosphere to enableremote sensing, weather data collection, telephony, and other functions. Geosynchronous (a.k.a. geostationary) orbits (GEO) lie at about 36,000 km altitude in the equatorial plane, where their 24-hour orbital period matches the rotation of the Earth. This orbit allows GEO satellites to hover above a fixed longitude, useful for communications and monitoring of large-scale weather patterns. Satellites in highly elliptical orbits (HEO) perform specialized functions inaccessible to other orbits. For example, HEO satellites in high inclination orbits provide wide-area communications above high-latitude regions for several hours at a time. **Figure 10-2** illustrates common orbits.

Line-of-Sight Exposure to a Nuclear Detonation

A nuclear device will, upon detonation, radiate a portion of its total yield as X-rays, with the fraction realized a function of weapon design and attached delivery system. Attenuation of X-rays propagating through the upper atmosphere is primarily by photo- electric absorption by oxygen and nitrogen and therefore is a function of X-ray spectrum, with higher-energy photons penetrating greater path-integrated mass density along the line of sight. Consequently, for a detonation above a (spectrally dependent) threshold altitude, X-rays emitted horizontally or upward will propagate to large distances virtually unattenuated by the atmosphere. X-rays emitted downward will be absorbed over ranges of tens of kilometers upon reaching sufficiently dense air.

Neutrons and gamma rays emitted by a detonation similarly propagate upward great distances into space for detonations above threshold altitudes. However, owing to scattering and absorption cross sections substantially smaller than X-ray photoelectric cross sections, major atmospheric attenuation of these energetic emissions occurs at altitudes below approximately 40 km.

For detonations up to a few hundred kilometers altitude, blast wave interactions between expanding weapon debris and the atmosphere may convert a majority of the kinetic yield of the weapon to ultraviolet (UV) photons. These photons propagate upward into space with little attenuation. UV photons emitted horizontally and downward are absorbed in the vicinity of the burst point to form the UV fireball. UV production for bursts above a few hundred kilometers declines rapidly, with precise values for these transition altitudes being functions of weapon output characteristics and dynamics. The combined flux of energetic photons (X-ray, gamma, and UV) and neutrons irradiates a vast region of space, diminished by spherical divergence, as shown in **figure 10-3**. The actual size of the hazard zone depends on weapon yield, detonation altitude, and the degree of satellite hardening against disruption or harm. Damage to satellite structures and to coatings on solar panels and sensor optics occurs when X-ray and UV fluxes exceed critical thresholds. Electronics damage similarly ensues when X-ray and gamma pulses induce destructive electric currents in circuit elements and when energetic neutrons penetrate solid-state circuitry.

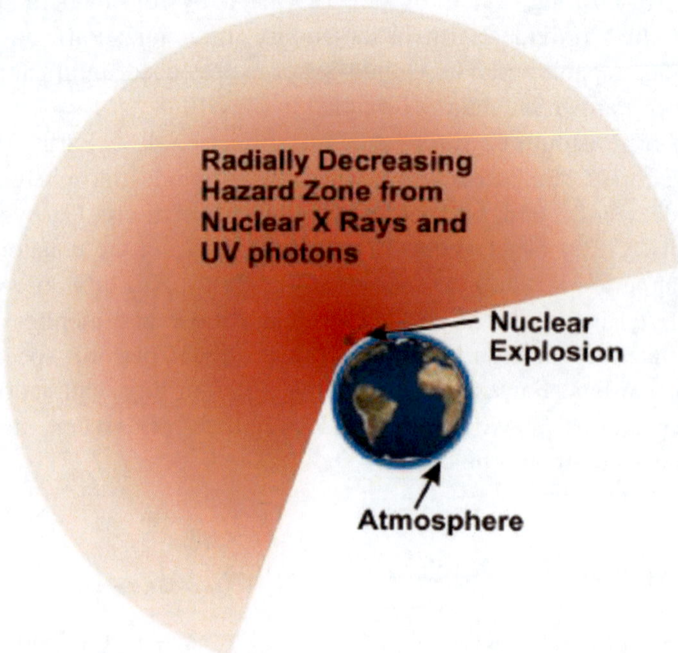

Where not shadowed by the Earth or shielded by atmospheric attenuation, X-rays and UV photons travel great distances from a high-altitude nuclear detonation where they may inflict damage to satellites.

Figure 10-3. Areas of Space Irradiated by Photons and Neutrons.

Persistently Trapped Radiation and Its Effects

In 1957, N. Christofilos at the University of California Lawrence Radiation Laboratory postulated that the Earth's magnetic field could act as a container to trap energetic electrons liberated by a high-altitude nuclear explosion to form a radiation belt that would encircle the Earth.[104] In 1958, J. Van Allen and colleagues at the State University of Iowa used data from the Explorer I and III satellites to discover the Earth's natural radiation belts.[105] **Figure 10-4** provides an idealized view of the Van Allen belts. Later in 1958, the United States conducted three low-yield ARGUS high-altitude nuclear tests, producing nuclear radiation belts detected by the Explorer IV satellite and other probes. In 1962, larger tests by the United States and the Soviet Union produced more pronounced and longer lasting radiation belts that caused deleterious effects to satellites then in orbit or launched soon thereafter.

A nuclear detonation is a significant source of free electrons originating from the highly ionized plasma that is a product of the nuclear blast. Nuclear detonations also create trapped radiation by beta decay of radioactive weapon debris and free-space decay of neutrons from the explosion, thereby creating electrons with energies up to several million electron volts (MeV). The most notable tests producing radiation hazards to satellites were the U.S. STARFISH detonation and three high-altitude tests by the Soviets, all conducted in 1962.

One assesses natural and trapped nuclear radiation effects on contemporary satellites by calculating repeated passage of a satellite through radiation belts over the satellite's life- time. While the geometry of a satellite's orbit is relatively straightforward, characterization of spatial and temporal properties of both natural and nuclear radiation belts is a complex problem. Nevertheless, one can establish relative scaling of levels of vulnerability from radiation belt geometry, as shown in **figure 10-5**. Intensities of radiation belts depend strongly on burst latitude. A burst at low latitude fills a small magnetic flux tube volume, so trapped flux tends to be concentrated and intense. The same burst at higher latitude fills a much larger magnetic flux tube volume.

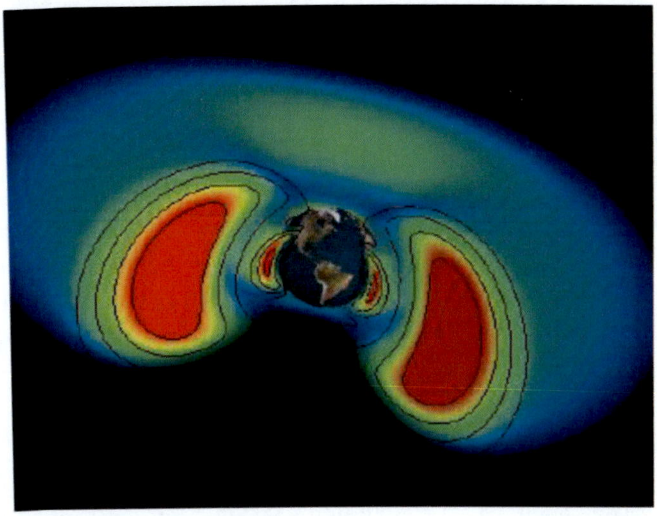

Figure 10-4. Naturally occurring belts (Van Allen belts) of energetic particles persistently trapped in the geomagnetic field are illustrated.

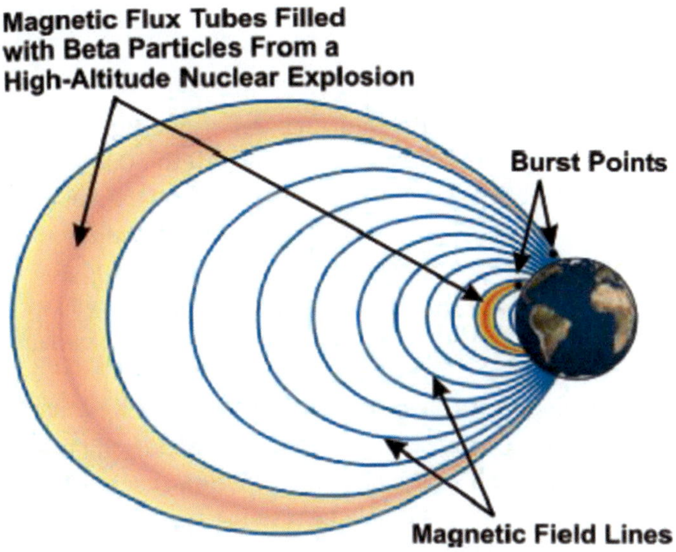

Figure 10-5. Schematic diagram of relative intensities of trapped fluxes from two identical high-altitude nuclear detonations.

All quantitative assessments of effects on satellite lifetime provided in this Section are based on calculations carried out using a code that tracks the satellite orbits through space and calculates the accumulated radiation dose.

Nuclear Weapon Effects on Electronic Systems

Electronic systems perform many critical spacecraft functions. An electronic power control system regulates the energy obtained from the solar cells. Attitude control circuits keep the vehicle oriented so that solar panels receive maximum exposure to the sun and sensors face the Earth. Information collected by sensors must be processed, stored, and transmitted to the Earth on demand. Communications satellites receive information, possibly process it, and then retransmit it, all by electronic circuits. Both prompt and long- term radiation effects have the potential for corrupting these functions in systems that lack hardening or other mitigation of nuclear effects.

Total-Dose Damage

A common criterion for failure of an electronic part is the total radiation energy per unit volume deposited in silicon. This absorbed energy density is expressed in rads(Si) (1 rad = 100 ergs/gram). Natural radiation to an electronic part in the International Space Station (ISS) behind a 2.54 mm semi-infinite (very large) aluminum slab averages about 100 rads per year. Previous literature has commonly used this shielding thickness for satellite radiation exposure calculations. However, it should be noted that electronics are placed in a variety of locations in a satellite and, therefore, can have different levels of shielding. Natural radiation to an electronic part in LEO, such as the National Oceanic and Atmospheric Administration (NOAA) satellite, in polar orbit behind a 2.54 mm semi-infinite aluminum slab is, on long-term average, about 620 rads per year, while some satellites with the same shielding might receive 50 kilorads per year.[106] Electronics must be shielded in accordance with the intended orbit to limit the dose received to a tolerable level.

Radiation-Induced Electrostatic Discharge

One hazard to spacecraft passing through the natural or nuclear radiation belts is internal or "deep dielectric" charging.[107] Lower-energy electrons (40 to 300 keV) become embedded in surface materials or poorly shielded internal materials and, on a timescale of hours to days, can build up sufficient electric field to cause a discharge, often resulting in satellite upset and occasionally in serious damage. Thermal blankets, external cables, and poorly shielded circuit boards are prime candidates for this type of charging. Modern coverglasses and optical solar reflectors are made sufficiently conductive to avoid such local charge buildup.

Radiation Effects Assessment and Hardening

Susceptibility of electronic components to nuclear weapon radiation has been studied intensively both experimentally and analytically since 1956. State-of-the-art computers and algorithms are used to extrapolate the experimental results to an operational environment.

The EMP Commission's mission was to evaluate the threat of high altitude nuclear weapon-induced EMP on American national infrastructure. A collateral result of a high altitude burst is a radiation threat to satellites, primarily those residing in LEO. The damage manifests as upset or burnout of sensitive microelectronics on the spacecraft. In some cases, damage can occur to external surfaces and structural members, as well as to optical components and to solar-cell power sources.

To address these issues, we considered a plausible set of 21 EMP nuclear events, which are listed in **table 10-1**. These disparate threats were then imposed upon a set of satellites (**table 10-2**) representative of the U.S. space infrastructure to examine the ancillary effects of an exoatmospheric nuclear detonation.

The time frame of interest is through the year 2015. As indicated in **table 10-1**, cases include both higher and lower yield weapons. Though not included in **tables 10-1** through **10-6**, each event is also associated with a particular latitude and longitude.

While the primary threat from nuclear-pumped radiation belts is to satellites in relatively low orbits, high-yield bursts could be detonated at latitudes and longitudes that would threaten higher orbiting satellites (Events 18 to 21). These bursts would be at relatively high latitudes sufficient to allow high-energy electrons to migrate along geomagnetic field lines that reach the high altitudes at which geosynchronous satellites reside.[109] Of course, at higher orbital altitudes, the density of ionizing radiation would be much reduced over that experienced by a satellite orbiting at lower altitudes and subjected to the same nuclear source due to the much larger volume in which the ionizing energy is distributed.

Table 10-1. Trial Nuclear Events

Event	Yield (kT)	Height of Burst (km)	L-Value[108]
1	20	200	1.26
2	100	175	1.09
3	300	155	1.09
4	10	300	1.19
5	100	170	1.16
6	800	368	1.27
7	800	491	1.36
8	4,500	102	1.11
9	4,500	248	1.16
10	30	500	1.23
11	100	200	1.18
12	20	150	1.24
13	100	120	1.26
14	500	120	1.26
15	100	200	1.03
16	500	200	1.03
17	5,000	200	1.03
18	1,000	300	4.11
19	10,000	90	4.19
20	1,000	350	6.85
21	10,000	90	6.47

Table 10-2. Analysis of Satellites

Satellite	Altitude (km)	Mission
NOAA/DMSP	800 (LEO)	Weather, remote sensing, search and rescue
TERRA/IKONOS	700 (LEO)	Moderate-high resolution imaging Earth resources and Earth sciences High resolution imagery, digital photography
ISS	322 (LEO)	Space science and technology
Generic GEO	GEO	Remote sensing
Generic HEO	HEO	Launch detection and other

It is emphasized that these events were chosen only for purposes of effects analysis. The satellites (**table 10-2**) were chosen to be representative of the many types and missions in orbit and to be representative targets for the radiation effects.

Prompt Radiation Effects

When a weapon is detonated at high altitude, satellites that lie within line of sight of the burst will be subject to direct (X-ray) radiation. Satellites in the shadow cast by the Earth will not be directly irradiated, as illustrated in **figure 10-3**, but will be subject to electron radiation as they transit debris and decay-products (primarily energetic beta electrons) mentioned previously that are trapped in the Earth's magnetic field. If there is a significant mass of intervening atmosphere between the detonation point and the satellite, direct nuclear radiation will be attenuated. Lacking this intervening shield, the radiation fluence will decrease as the inverse square of the distance.

Worst-case situations occur when a satellite is nearest the burst; for example, directly above or below it. In such cases, the range between satellite and burst is minimized, and X-ray, gamma, and neutron fluences on the satellite are maximized. Full evaluation of this hazard requires statistical analysis. The likelihood that the satellite will be in direct line of sight of the burst is typically 5 to 20 percent, depending on orbital parameters for the satellite and burst location. Even then, damage may be ameliorated by either distance or intervening atmosphere.

Calculations of X-ray exposure probabilities were performed for Events 9, 13, 17, and 18. The calculations yield the probability that a specific satellite will be exposed to a specified level of X-ray fluence. Results appear in **table 10-3**. With this information, one can estimate the probability of satellite damage based on known damage thresholds for spacecraft materials. Thresholds for various types of damage were chosen at, or close to, values accepted by the engineering community. Here, thermomechanical damage refers to removal or degradation of the coatings on solar cell surfaces. Depending on nuclear weapon output spectra, coating damage is generally a satellite's most sensitive thermomechanical damage mode. SGEMP (System-Generated EMP) burnout is damage caused by currents associated with X-ray-induced electron emission. Latch-up is a logic state setting of a semiconductor device that becomes frozen as a result of radiation exposure. Latch-up may cause large currents to flow in the affected circuit, resulting in unacceptable current-induced damage (i.e., burnout).

Line-of-sight exposure of the ISS to photons can cause significant damage to the solar-array coverglass coatings for Events 6, 7, 8, 9, and 17. NOAA/DMSP and TERRA/IKONOS are unlikely to be promptly affected thermomechanically by a line-of-sight photon exposure in any of our postulated nuclear events. Satellites in GEO are sufficiently far away because of their higher altitudes that the inverse square fall-off of the radiation reduces the potential exposure to a tolerable level.

Table 10-3. Probability That Satellites Suffer Damage by Direct Exposure to X-Rays

Satellite	Event	Probability of damage due to thermomechanical damage (%)	Probability of damage due to SGEMP/burnout (%)	Probability of damage due to latch-up/burnout (%)
ISS	9	1.7	4	4.2
	18	0	5	5
	13	~ 0	3	4
	17	1.7	5	5
NOAA	18	0.2	19	20
	13	0	3	5
	17	1	7	8
TERRA	18	~ 0.3	18	18
	13	0	2	5
	17	1.2	7	7

Permanent Damage from Exposure to the Enhanced Electron Belts

For this report, nuclear-enhanced electron belts are modeled as though they were providing a relatively constant trapped-electron environment. **Tables 10-4**, **10-5**, and **10-6** display reduced lifetimes of satellites that result from 17 of the 21 events. Results of events 18 through 21 will be discussed below.

Reduction in satellite lifetime is based on total dose from higher energy electrons to internal electronics, assumed to be shielded by a 0.100-inch slab of aluminum. In evaluating the biological response of astronauts to radiation on the ISS, 0.220 inches of slab shielding was assumed because the astronauts would usually be inside the pressurized modules of the space station. Some critical electronics for the station were still assumed to be shielded by only 100 mils of aluminum. Satellites are assumed to be hardened to twice the long-term-average natural background radiation encountered during a nominal mission.[110] Just as with photons, damage to spacecraft thermal, optical, and other surface coatings is caused by exposure to electrons of relatively low energies.

Except for the ISS in Event 4, even the low-yield events are capable of imposing a much-reduced lifetime on the satellites.

In the set of events depicted in **table 10-5**, the large weapon used in Event 17 inflicts severe damage on the ISS. Significantly, this exposure would cause radiation sickness to the astronauts within approximately 1 hour and a 90 percent probability of death within 2 to 3 hours.

Events 6 through 11 (**table 10-6**) were chosen within a geographical region where satellites could be placed at risk from a direct EMP attack resulting from regional contingencies.

Table 10-4. Trial Events in Group 1

Event	Yield (km)	HOB (kT)	Time to Failure (days)		
			NOAA	TERRA	ISS
1	20	200	30	70	150
2	100	175	15	30	50
3	300	155	4	7	9
4	10	300	20	60	5,400
5	100	170	30	70	100

Table 10-5. Trial Events in Group 2

Event	Yield (km)	HOB (kT)	Time to Failure (days)		
			NOAA	TERRA	ISS
12	20	150	25	60	230
13	100	120	60	200	200
14	500	120	4	6	3
15	100	200	10	20	30
16	500	200	1	3	4
17	5,000	200	0.1	0.1	0.1

Table 10-6. Trial Events in Group 3

Event	Yield (km)	HOB (kT)	Time to Failure (days)		
			NOAA	TERRA	ISS
6	800	368	1	1	0.5
7	800	491	1	1	1
8	4,500	102	0.1	0.2	0.2
9	4,500	248	0.1	0.2	0.2
10	30	500	40	100	150
11	100	200	10	17	20

The results of weapons detonated at high latitudes (Events 18 through 21) produced no dramatic nuclear effects. This is largely because these satellites are designed to operate in a far more hostile natural space environment due to the solar wind than are those in LEO.

Generally, most papers dealing with satellite lifetimes following a high-altitude nuclear detonation treat radiation effects on newly launched satellites with no pre-burst accumulated total dose. Except for satellites launched as replacements after a detonation, a more realistic assessment would assume a high-altitude detonation after a satellite had been in orbit for a portion of its anticipated service life. If a satellite is near the end of its design lifetime (i.e., has accumulated the majority of the total dose it can tolerate) prior to the detonation, the dose absorbed from a nuclear-pumped belt could cause prompt demise. To evaluate potential life-shortening effects on satellites, we examined a constellation of generic satellite systems. To assess sensitivity to assumed hardening level, we evaluated two hypothetical constellations. One constellation was assumed hardened to 1.5 times the natural total dose anticipated over the design lifetime (1.5x). The other constellation was assumed hardened to 2x. The scenario involved a 10 MT burst (50 percent fission yield) detonated on May 23, 2003, at an altitude of 90 km over northern Lake Superior (48.5 degrees north latitude, 87 degrees west longitude). Total dose for each constellation was based on realistic code calculations.

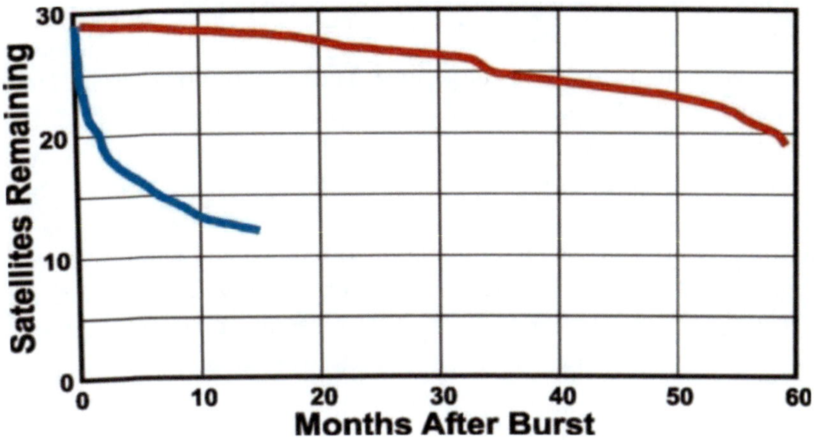

Figure 10-6. Satellites remaining after a 10 MT burst over Lake Superior.

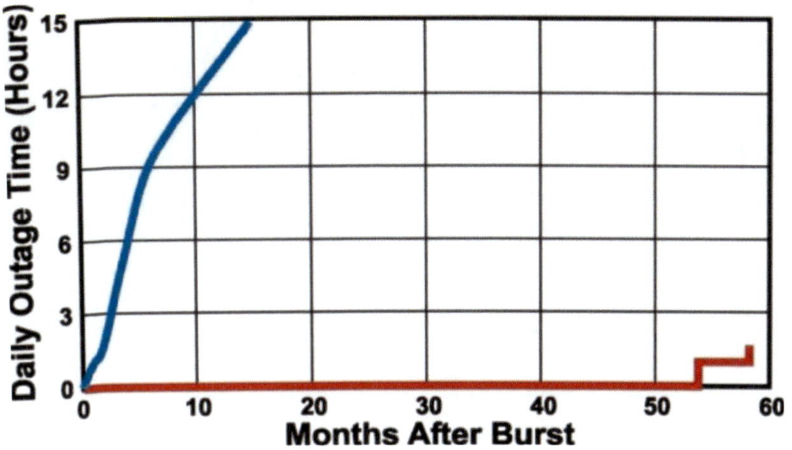

Figure 10-7. Satellite ground-based receiver outage time after a 10 MT burst over Lake Superior.

Figure 10-6 shows the resulting number of satellites remaining as a function of time after the burst. The blue and red curves correspond to the constellations hardened to 1.5x and 2x, respectively. Corresponding outage times for ground-based receivers are shown in **Figure 10-7**. Clearly, decreasing satellite hardening by 25 percent has a marked effect on survivability in this case.

HEO satellites already reside in orbits that are relatively challenging in terms of the natural radiation environment. Assuming these satellites are hardened to twice the natural dose they would normally accumulate in 15 years, a satellite's electronics would be hardened to approximately 325 krad behind a 100 mil (0.1 inch) semi-infinite slab of aluminum. With this level of hardness, one would expect that these satellites would not be vulnerable to a high-altitude burst of a single, low-yield (approximately 50 kT) device of unsophisticated design. Realistic code calculations suggest this is indeed the case.

Three large-yield events were investigated to determine whether they would present a threat to HEO satellites. Two of these events (Events 11 and 21) would not present a total ionizing dose problem for the satellite. Although Event 21 is a 10-MT burst, it has little effect on a HEO satellite because the trapped electrons are spread out over a large L-shell region. In contrast, the 100 kT of Event 11 does result in some detectable radiation accumulation on the satellite as it passes through altitudes near perigee. The yield is, however, too low to present a threat to the satellite. A 5-MT burst depicted in Event 17, on the other hand, does present a substantial threat to HEO satellites, given the hardening assumptions mentioned earlier. **Figure 10-8** shows that the assumed 2x natural hardening level of the satellite is exceeded about 36 days after Event 17.

Analysis of direct EMP attacks over the northern continental United States (CONUS) or Canada indicates lesser risk to LEO satellites from weapons with yields ranging from 10 kT to 100 kT. For yields approaching 1 MT (or greater) detonated at such latitudes, it becomes more difficult to predict the fate of LEO satellites. The larger yields make more severe nuclear-enhanced trapped flux environments, but depletion rates of trapped fluxes (both natural and nuclear) are difficult to predict.

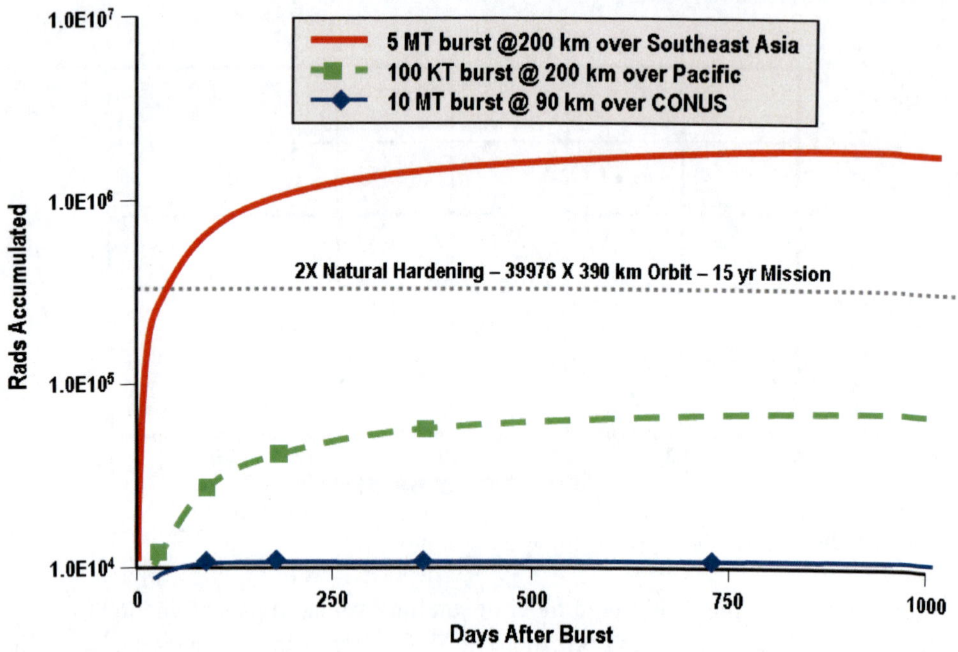

Figure 10-8. HEO satellite exposure to trapped radiation produced by Events 11, 17, and 21.

Satellite Ground Stations

Although bursts over CONUS may not directly damage satellites, the EMP effect on ground control stations could still render some satellites inoperable. We have focused our analyses on collateral weapon effects on satellites, without discussion of EMP effects on ground stations used for uplinks, downlinks, and satellite control. Currently, many of the important satellite systems use unique transmission protocols with dedicated ground terminals. Unique protocols limit interoperability, so loss of dedicated ground terminals could readily compromise overall functionality of a system, even if the system's satellites remained undamaged.

While satellites generally are designed to operate autonomously, with periodic housekeeping status downloads to ground controllers and uploads of commands, once damaged, satellites may require frequent, perhaps continuous, control from the ground to remain even partially functional. Thus, loss of ground stations to EMP could render otherwise functional satellites ineffective or lead to premature loss.

A comprehensive analysis of overall satellite system degradation should include potential loss of ground stations and cost/benefit trade-offs with respect to EMP hardening. A scenario-based analysis would reveal the extent to which loss of individual ground stations may pose an additional level of vulnerability.

Discussion of Results

Given inherent satellite fragility owing to severe weight constraints, any nation with missile lift capability and sufficient technology in the requisite disciplines can directly attack and destroy a satellite. Such attacks are outside the focus of this study. The Commission considered only hazards to satellites that may arise as collateral nuclear weapon effects during an EMP attack. The prominent collateral hazards are prompt nuclear output (X-rays, gamma rays, and neutrons), high fluences of UV photons generated by some high-altitude nuclear detonations, and nuclear burst enhancement (pumping) of the radiation belts surrounding the Earth in the region of space where satellites orbit.

The worst-case exposure of a satellite to direct x-radiation from a nuclear weapon can be lethal. For LEO satellites, the threat can be nonnegligible, but for satellites at GEO, the large distance between a detonation designed for an EMP attack and a satellite makes the probability of direct damage very low. The same argument holds for exposure to gamma rays, neutrons, and burst-generated UV light.

Nuclear-enhanced radiation belts must be considered differently, owing to their persistence and wide spatial distribution around the Earth. Because the natural trapped radiation environment at GEO is more severe on average than at most LEO locations, satellites at GEO typically are hardened to a greater extent than LEO satellites. Absent large yields (megatons), burst-generated energetic electron fluxes trapped in high-latitude (i.e., high L-shell) magnetic flux tubes generally are not sufficiently intense and long-lasting to cause the early demise of satellites in GEO, unless those satellites have accumulated sufficient natural radiation exposure to put them near the end of their service lives.[111]

Satellites in LEO are much more susceptible to damage from both direct and persistent radiation that results from an EMP attack, but the possibility of damage is highly dependent on weapon parameters (latitude and longitude, height of burst [HOB], and weapon yield).

Line-of-sight exposure of LEO research satellites such as ISS to X-ray and UV photons can cause significant damage to solar-array coverglass coatings for Events 6, 8, 9, 17, and 19. While such exposures are statistically infrequent, in those instances where they occur, they will result in immediate loss of many operational capabilities, as well as loss of power generating capacity.

The low-energy component of trapped-electron flux from beta decay of fission products and decay of free neutrons exceeds the long-term average natural flux for the high-yield Events 8, 9, and 17. Such flux levels will cause electrostatic breakdown in certain types of thermal radiator coatings and external cables on NOAA and TERRA within the first few days following the burst.

Uncertainties in Estimates

Uncertainties in satellite vulnerabilities result from imprecise knowledge of threat environments, combined with uncertainties in responses of satellite materials to those environments. Difficulties in characterizing aging effects of materials exposed to on-orbit conditions for extended periods exacerbate these uncertainties.

In the following comments, it is assumed that the weapons in question mirror U.S. technology available in the time frame 1970 to 1980.

Uncertainties in direct line-of-sight exposure of a satellite to radiation from a nuclear detonation result primarily from unknowns associated with the design of an offensive weapon, its delivery system, and its detonation altitude. These factors determine the fraction of weapon yield emitted as photons, neutrons, and beta particles and, hence, the type and magnitude of damage they inflict on satellites. Variability of weapon designs is estimated to lead to an uncertainty of approximately plus or minus a factor of five in UV hazard source strength (radiation primarily emitted from a weapon's case and its packaging within a delivery vehicle [but see below for more on UV photons]). Based on computational correlations with experimental data, there exits at least a factor of 10 uncertainty in X-ray spectral intensity at arbitrary photon energies of a kilovolt or more. Uncertainties in gamma-ray fluence and flux predictions are thought to be on the order of ±15 percent, as are those for prompt neutrons. Total yield is believed to be accurate to ±10 percent.

For bursts below a few hundred kilometers altitude, the debris-air blast-wave-generated fluence of UV photons (which can be as large as 80 percent of the kinetic yield of the device) carries an estimated uncertainty factor of 3 to 10, depending primarily on device characteristics. These uncertainty factors are ameliorated to some degree by decreasing burst altitude. Detonations below approximately 90 km occur in sufficiently dense air that UV photons are largely absorbed before they can escape to space.

Uncertainties in trapped radiation environments from high-altitude nuclear detonations also result from unknowns in offensive weapon design, but additional uncertainties arise in dispersal of radioactive weapon debris, efficiency with which beta particles become trapped in the geomagnetic field, subsequent transport of trapped particles, and the rapidity with which nuclear-burst enhancements of the radiation belts decay into the natural background. Under the best of circumstances, uncertainties in the intensity and persistence of trapped radiation estimated for the events considered in this report are at least a factor of 10 and are likely substantially more in situations that depart from limited circumstances of past nuclear tests.

Findings

Potential Vulnerabilities

An EMP attack on any of several important geographic regions could cause serious damage to LEO satellites. The STARFISH high-altitude nuclear burst greatly enhanced the high-energy electron environment in LEO, resulting in the early demise of several satellites on orbit at the time.[112] Copious documentation exists that describes recent radiation-induced satellite failures due to the natural radiation environment alone.

Given the large uncertainties discussed above, there may be a temptation to ignore the issue of high-altitude nuclear threats to satellites for the time being simply because insufficient information is available to implement a cost-effective protection solution. We believe that ignoring the issue would be ill advised for a number of reasons, including the consequences of losing possibly tens of billions of dollars in LEO space assets in a short time.

Mitigation of Threats

Any adversary possessing a lift and orbiting control capability can destroy a satellite: it is clearly neither cost effective nor desirable to harden every satellite against every possible threat. The challenge is to weigh risks/rewards of mitigation against mission priorities and plausible threats. A number of threat mitigation measures exist or have been proposed as an alternative or supplement to hardening.

Any combination of hardening and mitigation options can be chosen to achieve the required degree of survivability. Alternatives must be explored, documented, and reviewed so that management and users of space assets can make rational appraisals of the costs, benefits, and consequences of space system degradation and/or loss.

Hardening of Satellites and Ground Stations

Commercial satellites are hardened against their natural orbital environment to achieve the lifetime necessary to realize a profit. The technology to accomplish this goal is built into their design and factored into their cost. Protection from nuclear threats is not provided to commercial satellites because, from the commercial operator's perspective, it is not cost effective to do so.

The cost of hardening a system has been a subject of continuing controversy for the past 45 years. Systems project offices tend to estimate high to avoid the introduction of measures that threaten to escalate system cost. Achievable cost control is contingent upon ab initio design of radiation hardness into the system rather than on retrofitting it. Options other than hardening and shielding include repositioning selected satellites in times of stress to minimize exposure to enhanced radiation belts.

If the ground stations for satellites in any orbit are not hardened to EMP, the utility of the satellites could degrade, depending on their ability to operate autonomously.

Recommendations

- Each Federal Government organization that acquires and/or uses space should execute a systematic assessment of the significance of each such space system, particularly those in low Earth orbits, to its missions. Information from this assessment and associated cost and risk judgments will inform senior government decision-making regarding protection and performance assurance of these systems, so that each mission can be executed with the required degree of surety in the face of possible threats.

SECTION 11. GOVERNMENT

Introduction

A primary role of the Federal Government is to defend the Nation against threats to its security. EMP represents one such threat. Indeed, it is one of a small number of threats that can hold our society at risk of catastrophic consequences. The Executive branch of the Federal Government bears the responsibility for executing a strategy for dealing with this threat. The Commission has recommended a strategy for addressing this threat that combines prevention, protection, and recovery. It represents what we believe to be the best approach for addressing the EMP threat.

The Commission has identified an array of recommendations relating to civilian infrastructures that are logical outgrowths of our recommended strategy. Those recommendations relating to civilian infrastructures are contained in the individual Sections of this volume and will not be repeated here. Implementation of these recommendations will result in the identification of responsibilities at the regional, state, and local levels.

The Federal Government not only has the responsibility for being appropriately postured to cope with all aspects of the EMP threat, including preparations for recovery, but also has the responsibility to be able to respond to and manage national recovery in a competent and effective manner in the wake of an EMP attack. American citizens expect such competence and effectiveness from responsible government officials at all levels. In order to properly manage response and recovery, essential government functions will have to survive and function in the wake of an EMP attack.

Maintaining Government Connectivity and Coherence

It is essential that the Government continues to function through an electromagnetic pulse (EMP) emergency. Events over the last few years have highlighted the need for assured and real-time communications connectivity between government leadership and organizational assets for both crisis management and the management of a controlled recovery. Plans to ensure the continued functioning of government are embodied in Continuity of Operations (COOP) plans prepared by government organizations in anticipation of emergency situations and Continuity of Government (COG) planning to ensure the survival of constitutional government. National Security Presidential Directive 51 (NSPD 51) and Homeland Security Presidential Directive 20 (HSPD 20 on the subject of "National Continuity Policy", as described in a White House summary released May 9, 2007[113]), outlines these issues and directs the implementation of COOP and COG (excerpts noted below). The EMP

Commission met with National Security Council staff to discuss COG-related issues as they might relate to EMP threats. However, COG planning remains highly classified, and only this top-level overview can be provided within this venue.

Recommendations

- The Department of Homeland Security (DHS) should give priority to measures that ensure the President and other senior Federal officials can exercise informed leadership of the Nation in the aftermath of an EMP attack and that improve post-attack response capabilities at all levels of government.
- The President, Secretary of Homeland Security, and other senior officials must be able to manage national recovery in an informed and reliable manner. Current national capabilities were developed for Cold War scenarios in which it was imperative that the President have assured connectivity to strategic retaliatory forces. While this requirement is still important, there is a new need for considerably broader and robust connectivity between national leaders, government at all levels, and key organizations within each infrastructure sector so that the status of infrastructures can be assessed in a reliable and comprehensive manner and their recovery and reconstitution can be managed intelligently. The DHS, working through the Homeland Security Council, should give high priority to identifying and achieving the minimum level of robust connectivity needed for recovery following an EMP attack. In doing so, DHS should give particular emphasis to exercises that evaluate the robustness of the solutions being implemented.
- Working with state authorities and private sector organizations, the DHS should develop draft protocols for implementation by emergency and other government responses following an EMP attack, Red Team these extensively, and then institutionalize validated protocols through issuance of standards, training, and exercises.

NSPD 51/HSPD 20
Subject: "National Continuity Policy"
9 May 2007

Purpose

(1) This directive establishes a comprehensive national policy on the continuity of Federal Government structures and operations and a single National Continuity Coordinator responsible for coordinating the development and implementation of Federal continuity policies. This policy establishes "National Essential Functions," prescribes continuity requirements for all executive departments and agencies, and provides guidance for State, local, territorial, and tribal governments, and private sector organizations in order to ensure a comprehensive and integrated national continuity program that will enhance the credibility of our national security posture and enable a more rapid and effective response to and

recovery from a national emergency.

Definitions

(2) In this directive:

(a) "Category" refers to the categories of executive departments and agencies listed in Annex A to this directive;

(b) "Catastrophic Emergency" means any incident, regardless of location, that results in extraordinary levels of mass casualties, damage, or disruption severely affecting the U.S. population, infrastructure, environment, economy, or government functions;

(c) "Continuity of Government," or "COG," means a coordinated effort within the Federal Government's executive branch to ensure that National Essential Functions continue to be performed during a Catastrophic Emergency;

(d) "Continuity of Operations," or "COOP," means an effort within individual executive departments and agencies to ensure that Primary Mission-Essential Functions continue to be performed during a wide range of emergencies, including localized acts of nature, accidents, and technological or attack-related emergencies;

(e) "Enduring Constitutional Government," or "ECG," means a cooperative effort among the executive, legislative, and judicial branches of the Federal Government, coordinated by the President, as a matter of comity with respect to the legislative and judicial branches and with proper respect for the constitutional separation of powers among the branches, to preserve the constitutional framework under which the Nation is governed and the capability of all three branches of government to execute constitutional responsibilities and provide for orderly succession, appropriate transition of leadership, and interoperability and support of the National Essential Functions during a catastrophic emergency;

(f) "Executive Departments and Agencies" means the executive departments enumerated in 5 U.S.C. 101, independent establishments as defined by 5 U.S.C. 104(1), Government corporations as defined by 5 U.S.C. 103(1), and the United States Postal Service;

(g) "Government Functions" means the collective functions of the heads of executive departments and agencies as defined by statute, regulation, presidential direction, or other legal authority, and the functions of the legislative and judicial branches;

(h) "National Essential Functions," or "NEFs," means that subset of Government Functions that are necessary to lead and sustain the Nation during a catastrophic emergency and that, therefore, must be supported through COOP and COG capabilities; and

(i) "Primary Mission Essential Functions," or "PMEFs," means those Government Functions that must be performed in order to support or implement the performance of NEFs before, during, and in the aftermath of an emergency.

Policy

(3) It is the policy of the United States to maintain a comprehensive and effective

continuity capability composed of Continuity of Operations and Continuity of Government programs in order to ensure the preservation of our form of government under the Constitution and the continuing performance of National Essential Functions under all conditions.

Implementation Actions

(4) Continuity requirements shall be incorporated into daily operations of all executive departments and agencies. As a result of the asymmetric threat environment, adequate warning of potential emergencies that could pose a significant risk to the homeland might not be available, and therefore all continuity planning shall be based on the assumption that no such warning will be received.

Emphasis will be placed upon geographic dispersion of leadership, staff, and infrastructure in order to increase survivability and maintain uninterrupted Government Functions. Risk management principles shall be applied to ensure that appropriate operational readiness decisions are based on the probability of an attack or other incident and its consequences.

...

(10) Federal Government COOP, COG, and ECG plans and operations shall be appropriately integrated with the emergency plans and capabilities of State, local, territorial, and tribal governments, and private sector owners and operators of critical infrastructure, as appropriate, in order to promote interoperability and to prevent redundancies and conflicting lines of authority. The Secretary of Homeland Security shall coordinate the integration of Federal continuity plans and operations with State, local, territorial, and tribal governments, and private sector owners and operators of critical infrastructure, as appropriate, in order to provide for the delivery of essential services during an emergency.

(11) Continuity requirements for the Executive Office of the President (EOP) and executive departments and agencies shall include the following:

(a) The continuation of the performance of PMEFs during any emergency must be for a period up to 30 days or until normal operations can be resumed, and the capability to be fully operational at alternate sites as soon as possible after the occurrence of an emergency, but not later than 12 hours after COOP activation;

(b) Succession orders and pre-planned devolution of authorities that ensure the emergency delegation of authority must be planned and documented in advance in accordance with applicable law;

(c) Vital resources, facilities, and records must be safeguarded, and official access to them must be provided;

(d) Provision must be made for the acquisition of the resources necessary for continuity operations on an emergency basis;

(e) Provision must be made for the availability and redundancy of critical communications capabilities at alternate sites in order to support connectivity between and among key government leadership, internal elements, other executive departments and agencies, critical partners, and the public;

(f) Provision must be made for reconstitution capabilities that allow for recovery from a catastrophic emergency and resumption of normal operations; and

(g) Provision must be made for the identification, training, and preparedness of personnel capable of relocating to alternate facilities to support the continuation of the performance of PMEFs.

...

(19)Heads of executive departments and agencies shall execute their respective department or agency COOP plans in response to a localized emergency and shall:

(a) Appoint a senior accountable official, at the Assistant Secretary level, as the Continuity Coordinator for the department or agency;

(b) Identify and submit to the National Continuity Coordinator the list of PMEFs for the department or agency and develop continuity plans in support of the NEFs and the continuation of essential functions under all conditions;

(c) Plan, program, and budget for continuity capabilities consistent with this directive;

(d) Plan, conduct, and support annual tests and training, in consultation with the Secretary of Homeland Security, in order to evaluate program readiness and ensure adequacy and viability of continuity plans and communications systems; and

(e) Support other continuity requirements, as assigned by category, in accordance with the nature and characteristics of its national security roles and responsibilities

...

GEORGE W. BUSH

SECTION 12. KEEPING THE CITIZENRY INFORMED: EFFECTS ON PEOPLE

Introduction

The best current estimate is that the electromagnetic pulse (EMP) produced by a high-altitude nuclear detonation is not likely to have direct adverse effects on people. Such effects have not been observed for the personnel who operate EMP simulators.[114] Medical surveillance studies on human exposure to pulsed electromagnetic fields have supported this inference.[115]

An important exception is people whose well-being depends on electronic life support equipment. They will be directly impacted by effects that disrupt or damage such devices. Research sponsored by the Commission suggests that some heart pacemakers may be among the devices susceptible to upset from high-altitude EMP.[116,117]

While most effects on people would be indirect, they could be significant in a just-in-time economy in which local stocks of medicines, baby food, and other health-critical items are limited. The physical consequences of the serious high-altitude EMP attacks on the United States (U.S.) of concern to the Commission would likely include the failure of the electric power grid and degradation of telecommunication systems, computers, and electronic components over large areas of the country. A disruption of this scale could cripple critical infrastructures and hinder the delivery of day-to-day necessities, because of the interconnectivity of telecommunication networks and the electrical dependence of most cities, government agencies, businesses, households, and individuals. It also could require a long recovery period. To assess human consequences, the contingency of concern is one in which electricity, telecommunications, and electronics are out of service over a significant area for an extended period of time.

The human consequences of such a scenario include the social and psychological reactions to a sudden loss of stability in the modern infrastructure over a large area of the country. Loss of connectivity between the government and its populace would only exacerbate the consequences of such a scenario.

This analysis is based largely on selected case studies, including major blackouts, natural disasters, and terrorist incidents in recent U.S. history. These incidents served as approximate analogs in order to best predict the sociological and psychological effects of an EMP attack.

Impact of an EMP Attack

While no single event serves as a model for an EMP scenario with incidence of long-lasting widespread power outage, communications failure, and other effects, the combined analysis of the following case studies provides useful insight in determining human reactions following an EMP attack:

Blackouts:

- Northeast (1965)
- New York (1977)
- Hydro Quebec (1989)
- Western states (1996)
- Auckland, New Zealand (1998)
- Northeast (2003)

Natural Disasters:

- Hurricane Hugo (1989)
- Hurricane Andrew (1992)
- Midwest floods (1993)

Terrorist Incidents:

- World Trade Center attack (2001)
- Anthrax attacks (2001)

Blackouts

In 1965, a blackout occurred over the northeastern United States and parts of Canada. New Hampshire; Vermont; Massachusetts; Connecticut; Rhode Island; New York, including metropolitan New York City; and a small part of Pennsylvania were in the dark after operators at Consolidated Edison were forced to shut down its generators to avoid damage. Street traffic was chaotic, and some people were trapped in elevators, but there were few instances of antisocial behavior while the lights were out.[118] It was a "long night in the dark," but the recovery proceeded without incident, and citizens experienced relative civility.

TIME Magazine described New York's next blackout, in 1977, as a "Night of Terror." [119] Widespread chaos reigned in the city until power was restored — entire blocks were looted and set ablaze, people flipped over cars and vans on the streets; the city was in pandemonium. That night 3,776 arrests were made, and certainly not all looters, thieves, and arsonists were apprehended or arrested.[120] While this is a dramatic example of antisocial behavior following a blackout, sociologists point to extraordinary demographic and historical issues that contributed to the looting. For instance, extreme poverty and socio-economic inequality plagued New York neighborhoods, and many of the looters originated from the poorer sections of the city, engaging in "vigilante redistribution" by looting consumer goods and luxuries. Racial tensions were high, and a serial killer known as Son of Sam had recently terrorized New Yorkers.

In 1989, more than 6 million customers lost power when the geomagnetic storm discussed in Section 4 caused a massive power failure in Quebec. The electricity failures caused by this geomagnetic storm reached a much larger area than is typically affected by traditional blackouts resulting from technological failure. However, the outage lasted just over 9 hours, most of which were during the day.[121] The local and national papers were curiously silent about the blackout, and little to no unusual or adverse human behavior was attributed to the power loss. The event was most significantly a lesson for operators of the North American electric grids because it revealed vulnerabilities in the system.

In 1998, Auckland, New Zealand, experienced a significant blackout that lasted more than 5 weeks and affected more than 1 million people.[122] Civility reigned for the duration of the outage, which was likely attributed to a number of factors, including:

- There was no significant threat to public health, because water and sewage infrastructures were functioning.
- In anticipation of potential incidents, police increased their presence in urban areas.
- The recovery process was underway nearly immediately, communicating to the public that the situation would eventually be under control.
- Nearly all blackout recovery resources of New Zealand were rushed to the capital for recovery efforts.

Recovery efforts from elsewhere in New Zealand were significant symbolically as well as practically, as demonstrated by the fact that electricity was available elsewhere. Businesses attempted to carry on as normally as possible, with some examples of opportunism, such as businesses relocating to more desirable spaces that had been vacated. Social consequences included criticism and blame of the authorities, both municipal and national, because the technological failures were attributed in large part to privatization of the power sector. However, this response never materialized into violence, crime, or social disorder.

Most recently, New York City and the eight states in the northeast experienced another significant blackout in August 2003. While the blackout inconvenienced many on a hot summer day, general civility remained intact. News coverage indicated that those affected by the blackout dealt with the obstacles quietly and even developed a sort of camaraderie while struggling through nights without running water and electricity. In contrast to the 1977 blackout, police made only 850 arrests the night of the 2003 black- out, of which "only 250 to 300 were directly attributable to the blackout," indicating a slight decline from the average number of arrests on a given summer day.[123] While this blackout was widespread, it was not long lasting, and it did not interrupt the communications infrastructure significantly.

Blackouts provide only a partial picture of life following an EMP attack. Most blackouts are localized and are resolved quickly. Further, usually communication systems are not completely shut down, and major infrastructures can remain intact if significant portions of infrastructure hardware are located outside of the affected area. In order to best approximate the effects of longer-lasting, widespread infrastructure disruption—with or without electrical power failure—it is necessary to look to natural disasters for examples of human reaction.

Natural Disasters

At the time that Hurricane Hugo hit in 1989, it was the most intense hurricane to strike Georgia and the Carolinas in 100 years. Surveys of Hurricane Hugo's survivors indicate that some individuals who suffered personal and financial losses from the hurricane showed clinically significant symptoms of psychological trauma. According to some researchers, many of the adverse mental health effects of Hugo could be explained by deterioration in perceived social support. While on the whole, the rate of post-traumatic stress disorder symptoms was low, stress effects lingered long after the hurricane's physical damage was repaired.

Hurricane Andrew blew through the southeastern United States and along the coast of the Gulf of Mexico in 1992, causing $26.5 billion in damage. Andrew left 250,000 families homeless and 1.4 million families without electricity immediately following the hurricane. After such extraordinary destruction and disruption, it is perhaps not surprising that one-third of a sample of individuals met criteria for post-traumatic stress disorder 4 months after the hurricane.[124]

Hurricanes Hugo and Andrew demonstrated to psychologists that disaster-related declines in perceived support explained the difference in symptoms between the two disasters; deterioration was more significant in Andrew and recovery was weaker. In the long-lasting recovery period, Floridians saw looting, opportunism, and vigilante civil defense. Press coverage of Hurricane Andrew suggests that after a multi-state disaster, people will expect help, and they will expect it from the federal government, as well as from state and local authorities.

Flooding in the American Midwest in 1993 resulted in 25 deaths, affected more than 8 million acres, and cost billions of dollars in property damage and more than 2 billion dollars in crop damage. Water depths ranged from 11 feet of flooding in Minneapolis to 43 feet in St. Louis. Electricity was restored where possible within 3 days and in down- town Des Moines within 23 hours. The floods devastated families, businesses, and individuals, who lost nearly everything and were unable to control events throughout the recovery process. Thousands of people assisted in volunteer recovery efforts by sandbagging and providing needed supplies.[125] Most came from unaffected areas to help the most urgent victims. The floods provide an example of widespread damage crippling several infrastructures for a significant period of time and an example of a disaster in which regional experience may matter tremendously in disaster recovery.

Blackouts and natural disasters have limits as approximations of recovery following an EMP attack. An important element is the relevance of fear and individual panic in these situations versus what might occur following an EMP attack. For this component, it is useful to examine recent terrorist incidents in the United States in order to gauge the effects of fear among the public. Because terrorist attacks appear to be indiscriminate and random, they can arouse acute anxiety and feelings of helplessness, which shatter beliefs of invulnerability and even a belief in justice and order in the world.

Terrorist Incidents

The attacks on the World Trade Center in New York on September 11, 2001, certainly qualified as seemingly indiscriminate and random. Following this disaster, in which nearly 3,000 people died, those in the immediate and surrounding area showed considerable psychological trauma and damage. Some individuals who experienced these attacks may have lost confidence in their abilities to cope and control outcomes. Overall, however, the survivors of the attacks proved remarkably resilient, flexible, and competent in the face of an arbitrary, violent, and completely unexpected attack.[126]

In October 2001, a month following the attack on the World Trade Center, Americans saw a series of anthrax-infected mail pieces threatening intended mail recipients and handlers. The death toll was small (five individuals), but public concern was considerable. This period is an example of public response to an adversary-initiated threat that disrupted infrastructure. The public demonstrated a great need for control over the situation, through preparedness and information. For example, many Americans took protective measures, despite the astronomical odds against infection. The news media were saturated with reports of anthrax infections, suspected infections, and general information about anthrax and how to respond to infection. Though no culprit was apprehended, the attacks stopped, and normal postal activity resumed.

Some Lessons Learned

Though the United States has not experienced a severe, widespread disruption to infrastructure comparable to an EMP attack, the cases reviewed provide some practical direction for predictions of behavior. For example, it can be expected that emotional reactions such as shock and paralysis that have followed past disasters could be magnified in a large-scale event such as an EMP attack. In particular, the paralysis of government assistance entities, such as law enforcement and emergency services, would aggravate this effect. In most instances, social disorder would be minimal, in significant part, due to the knowledge that authorities are in control of the situation. Without that assurance from an outside source, it appears likely that people would turn to immediate neighbors or community members for information and support, if possible.

Following disruptive disasters, information is among the most pressing needs for individuals. Not surprisingly, people's first concerns are the whereabouts and safety of their family members and friends. Another urgent priority is an understanding of the situation — knowledge of what has happened, who and what is affected, and the cause of the situation. A related yet distinct information need is for confirmation that the situation will be resolved, either from common sense and experience, in the case of a small-scale disaster, or from the involvement of local or federal authorities, in the case of a large-scale disaster. Psychologists note that dramatic events force people to reexamine their basic understanding about the world, and that survivors need to process an event before they can fully absorb it. This information processing begins the alternating phases of intrusion and avoidance that are primary indicators of post-traumatic stress.[127]

The aftermath of natural disasters is often marked by instances or a period of considerable pro-social behavior such as cooperation, social solidarity, and acts of selflessness. However, this encouraging observation might not be similarly magnified in projections for human behavior following an EMP attack. The key intangible, immeasurable difference is the knowledge that normal order would resume, based on significant indicators.

It is important to note some of the differences between natural disasters and technological disasters, particularly those caused by human intent. Natural disasters "create a social context marked by an initial overwhelming consensus regarding priorities and the allocation of resources,"[128] which explains the enormous outpouring of voluntary support following the floods of 1993. In contrast to natural disasters, which "occur as purpose-less, asocial events; civil disturbances can be viewed as instrumentally initiated to achieve certain social goals."[129] An EMP attack would certainly be perceived similarly, whether the adversary were a terrorist organization or a state.

The selected case studies provide only an approximation of EMP effects. For example, the effects of the knowledge that widespread infrastructure disruption resulted from an intentional foreign attack are yet unknown. Much evidence points to people's resilience in the immediate aftermath of disasters. However, during a lengthy recovery process, as would be expected following an EMP attack with widespread, long-duration effects, the psychological effects of the attack should not be underestimated.

It appears clear that the most crucial question in the task of avoiding social disorder is how to establish communication without electricity immediately following an EMP attack. Without communication alternatives, it would be impossible to alert people to the availability of emergency supplies or inform them concerning emergency response activities. It also appears clear that greater awareness of the nature of an EMP attack and knowledge of what prudent preparations might be undertaken to mitigate its consequences would be desirable. Accordingly we make the following recommendations.

Recommendations

- Support to national leadership should involve measures to ensure that the President can communicate effectively with the citizenry.
- Because many citizens would be without power, communications, and other services for days — or perhaps substantially longer — before full recovery could occur, during that interval, it will be crucial to provide a reliable channel of information to citizens to let them know what has happened, what the current situation is, when help of what types might be available, what their governments are doing, and answers to the host of other questions that, if not answered, would almost certainly create more instability and suffering for the affected individuals, communities, and the Nation as a whole. In particular:

- The Department of Homeland Security should play a leading role in spreading knowledge of the nature of prudent mitigation preparations for EMP attack to mitigate its consequences.
- The Department of Homeland Security should add content to Web sites it maintains, such as www.Ready.gov, which provides concise overviews of the threats posed by EMP attacks and geomagnetic storms, summarizes steps that people should take given an incident and identifies alternate or emergency communications channels.
- The Department of Homeland Security should work with state homeland security organizations to develop and exercise communications networks involving the organizations that normally operate in each community.

APPENDIX A. THE COMMISSION AND ITS CHARTER

The Commission to Assess the Threat to the United States from Electromagnetic Pulse (EMP) Attack was established by Congress through Title XIV of Public Law 106-398. Looking out 15 years, the Commission was tasked to assess:

1) The nature and magnitude of potential high-altitude EMP threats to the United States from all potentially hostile states or non-state actors that have or could acquire nuclear weapons and ballistic missiles enabling them to perform a high-altitude EMP attack against the United States within the next 15 years.
2) The vulnerability of United States military and especially civilian systems to an EMP attack, giving special attention to vulnerability of the civilian infrastructure as a matter of emergency preparedness.
3) The capability of the United States to repair and recover from damage inflicted on United States military and civilian systems by an EMP attack.
4) The feasibility and cost of hardening select military and civilian systems against EMP attack.

The Commission was also tasked to recommend any steps it believes should be taken by the United States to better protect its military and civilian systems from EMP attack.

In accord with its charter, the Commission focused on the electromagnetic pulse produced by high-altitude nuclear weapon detonations, as opposed to other types of nuclear and non-nuclear EMP phenomena. Unless clearly indicated to the contrary, all references to EMP are to the electromagnetic pulse produced by a high-altitude nuclear detonation.

This report presents the unanimous conclusions and recommendations of the Commissioners.

Organization

Commissioners were nominated by the Secretary of Defense and by the Administrator of the Federal Emergency Management Agency[130]:

- Dr. William R. Graham (Chairman)
- Dr. John S. Foster, Jr.
- Mr. Earl Gjelde
- Dr. Robert J. Hermann
- Mr. Henry (Hank) M. Kluepfel
- Gen Richard L. Lawson, USAF (Ret.)
- Dr. Gordon K. Soper
- Dr. Lowell J. Wood, Jr.
- Dr. Joan B. Woodard

Commissioners brought to this task a wide range of expertise, including service as an advisor to the President; senior management experience in both civilian and military agencies, National Laboratories, and the corporate sector; management and operation of national infrastructures, and technical expertise in the design of nuclear weapons and in the hardening of systems against nuclear weapon effects. Commissioner resumes are provided in an appendix to this volume.

Dr. Michael J. Frankel served as Executive Director of the Commission. He was also responsible for overseeing the technical efforts in support of the Commission accomplished by both American and foreign organizations. The Institute for Defense Analysis, under the leadership of Dr. Rob Mahoney, provided staff and facilities support for the Commission. Dr. Peter Pry provided liaison with the Congress. The Commission also benefited from the understanding of EMP available in foreign institutions. Several government, non-profit, and commercial organizations conducted work and prepared reports for the Commission.

Method

The Commission employed a capabilities-based methodology to assess potential high-altitude EMP threats to the United States over the next 15 years.[131] To this end it engaged the current Intelligence Community, sponsored the acquisition of new test data and performed analytic studies as input to the independent assessment developed by the Commission. Fifteen years is a very long time horizon. Many developments are possible, to include actions by the United States and others that can shape this future in a variety of ways. At the Commission's inception, Iraq was a state of concern from the standpoint of nuclear proliferation and potential EMP threats. Due to actions taken by the Coalition, such Iraqi capabilities are no longer a current concern.

> *...a tendency in our planning to confuse the unfamiliar with the improbable. The contingency we have not considered looks strange; what looks strange is therefore improbable; what seems improbable need not be considered seriously.*
>
> — Thomas C. Schelling, in Roberta Wohlstetter, Pearl Harbor: Warning and Decision. Stanford University Press, 1962, p. vii

The Commission did not attempt to forecast the relative likelihood of alternative WMD threat scenarios. Instead, it sponsored research and reviewed existing assessments to identify the capabilities that might be available to adversaries, with particular emphasis on ballistic missile and nuclear weapons needed for EMP attacks.

The Commission's charter encompassed all types of high-altitude EMP threats. The Commission made a decision to focus most of its efforts on the most feasible of these threats – EMP attacks involving one or a few weapons that could cause serious damage to the functioning of the United States as a society or result in undermining national support to American forces during a regional contingency.

Activities

The Commission received excellent support from the Intelligence Community, particularly the Central Intelligence Agency, Defense Intelligence Agency, National Security Agency, and Department of Energy Office of Intelligence. National Nuclear Security Administration laboratories (Lawrence Livermore, Los Alamos, and Sandia), the Navy, and the Defense Threat Reduction Agency provided excellent technical support to the Commission's analyses. While it benefited from these inputs, the Commission developed an independent assessment, and is solely responsible for the content of its research, conclusions, and recommendations in this report.

The Commission also reviewed relevant foreign research and programs, and assessed foreign perspectives on EMP attacks.

In considering EMP, the Commission also gave attention to the coincident nuclear effects that would result from a high-altitude detonation that produces EMP, e.g., possible disruption of the operations of, or damage to, satellites in a range of orbits around the Earth.

In addition to examining potential threats, the Commission was charged to assess U.S. vulnerabilities (civilian and military) to EMP and to recommend measures to counter EMP threats. For these purposes, the Commission reviewed research and best practices within the United States and other countries.

Early in this review it became apparent that only limited EMP vulnerability testing had been accomplished for modern electronic systems and components. To partially remedy this deficit, the Commission sponsored illustrative testing of current systems and infrastructure components.

Appendix B. Biographies

Dr. William R. Graham is Chairman of the Commission to Assess the Threat to the United States from Electromagnetic Pulse Attack. He is the retired Chairman of the Board and Chief Executive Officer of National Security Research Inc. (NSR), a Washington-based company that conducted technical, operational, and policy research and analysis related to U.S. national security. He currently serves as a member of the Department of Defense's Defense Science Board and the National Academies Board on Army Science and Technology. In the recent past he has served as a member of several high-level study groups, including the Department of Defense Transformation Study Group, the Commission to Assess United States National Security Space Management and Organization, and the

Commission to Assess the Ballistic Missile Threat to the United States. From 1986–89 Dr. Graham was the director of the White House Office of Science and Technology Policy, while serving concurrently as Science Advisor to President Reagan, Chairman of the Federal Joint Telecommunications Resources Board, and a member of the President's Arms Control Experts Group.

Dr. John S. Foster, Jr., is Chairman of the Board of GKN Aerospace Transparency Systems, and consultant to Northrop Grumman Corporation, Technology Strategies & Alliances, Sikorsky Aircraft Corp., Intellectual Ventures, Lawrence Livermore National Lab, Ninesigma, and Defense Group. He retired from TRW as Vice President, Science and Technology, in 1988 and continued to serve on the Board of Directors of TRW from 1988 to 1994. Dr. Foster was Director of Defense Research and Engineering for the Department of Defense from 1965–1973, serving under both Democratic and Republican administrations. In other distinguished service, Dr. Foster has been on the Air Force Scientific Advisory Board, the Army Scientific Advisory Panel, and the Ballistic Missile Defense Advisory Committee, Advanced Research Projects Agency. Until 1965, he was a panel consultant to the President's Science Advisory Committee, and from 1973–1990 he was a member of the President's Foreign Intelligence Advisory Board. He is a member of the Defense Science Board, which he chaired from January 1990–June 1993. From 1952–1962, Dr. Foster was with Lawrence Livermore National Laboratory (LLNL), where he began as a Division Leader in experimental physics, became Associate Director in 1958, and became Director of LLNL and Associate Director of the Lawrence Berkeley National Laboratory in 1961.

Mr. Earl Gjelde is the President and Chief Executive Officer of Summit Power Group Inc., and several affiliated companies, primary participants in the development of over 5,000 megawatts of natural gas fired electric and wind generating plants within the United States. He has served on the boards of EPRI and the U.S. Energy Association among others. He has held a number of U.S.A. government posts, serving as President George Herbert Walker Bush's Under (now called Deputy) Secretary and Chief Operating Officer of the U.S. Department of the Interior (1989) and serving President Ronald Reagan as Under Secretary and Chief Operating Officer of the U.S. Department of the Interior (1985–1988), the Counselor to the Secretary and Chief Operating Officer of the U.S. Department of Energy (1982–1985); and Deputy Administrator, Power Manager and Chief Operating Officer of the Bonneville Power Administration (1980–1982). While in the Reagan Administration he served concurrently as Special Envoy to China (1987), Deputy Chief of Mission for the U.S.-Japan Science and Technology Treaty (1987–1988), and Counselor for Policy to the Director of the National Critical Materials Council (1986–1988). Prior to 1980, he was a Principal Officer of the Bonneville Power Administration.

Dr. Robert J. Hermann is a Senior Partner of Global Technology Partners, LLC, a consulting firm that focuses on technology, defense aerospace, and related businesses worldwide. In 1998, Dr. Hermann retired from United Technologies Corporation (UTC), where he was Senior Vice President, Science and Technology. Prior to joining UTC in 1982, Dr. Hermann served 20 years with the National Security Agency with assignments in research and development, operations, and NATO. In 1977, he was appointed Principal Deputy Assistant Secretary of Defense for Communications, Command, Control, and Intelligence. In 1979, he was named Assistant Secretary of the Air Force for Research, Development, and Logistics and concurrently was Director of the National Reconnaissance Office.

Mr. Henry (Hank) M. Kluepfel is a Vice President for Corporate Development at SAIC. He is the company's leading cyberspace security advisor to the President's National Security Telecommunications Advisory Committee (NSTAC) and the Network Reliability and Interoperability Council (NRIC). Mr. Kluepfel is widely recognized for his 30- plus years of experience in security technology research, design, tools, forensics, risk reduction, education, and awareness, and he is the author of industry's de facto standard security base guideline for the Signaling System Number 7(SS7) networks connecting and controlling the world's public telecommunications networks. In past affiliations with Telcordia Technologies (formerly Bellcore), AT&T, BellSouth and Bell Labs, he led industry efforts to protect, detect, contain, and mitigate electronic and physical intrusions and led the industry's understanding of the need to balance technical, legal, and policy- based countermeasures to the then emerging hacker threat. He is recognized as a Certified Protection Professional by the American Society of Industrial Security and is a Senior Member of the Institute of Electrical and Electronics Engineers (IEEE).

General Richard L. Lawson, USAF (Ret.), is Chairman of Energy, Environment and Security Group, Ltd., and former President and CEO of the National Mining Association. He also serves as Vice Chairman of the Atlantic Council of the U.S.; Chairman of the Energy Policy Committee of the U.S. Energy Association; Chairman of the United States delegation to the World Mining Congress; and Chairman of the International Committee for Coal Research. Active duty positions included serving as Military Assistant to the President; Commander, 8th Air Force; Chief of Staff, Supreme Headquarters Allied Powers Europe; Director for Plans and Policy, Joint Chiefs of Staff; Deputy Director of Operations, Headquarters U.S. Air Force; and Deputy Commander in Chief, U.S. European Command.

Dr. Gordon K. Soper is employed by Defense Group Inc. There he has held various senior positions where he was responsible for broad direction of corporate goals relating to company support of government customers in areas of countering the proliferation of weapons of mass destruction, nuclear weapons effects and development of new business areas and growth of technical staff. He provides senior-level technical support on a range of task areas to the Defense Threat Reduction Agency (DTRA) and to a series of Special Programs for the Office of the Secretary of Defense and the White House Military Office. Previously, Dr. Soper was Principal Deputy to the Assistant to the Secretary of Defense for Nuclear, Chemical and Biological Defense Programs (ATSD(NCB)); Director, Office of Strategic and Theater Nuclear Forces Command, Control and Communications (C3) of the Office of the Assistant Secretary of Defense (C3I); Associate Director for Engineering and Technology/Chief Scientist at the Defense Communications Agency (now DISA); and held various leadership positions at the Defense Nuclear Agency (now DTRA).

Dr. Lowell L. Wood, Jr., is a scientist-technologist who has contributed to technical aspects of national defense, especially defense against missile attack, as well as to controlled thermonuclear fusion, laser science and applications, optical and underwater communications, very high-performance computing and digital computer-based physical modeling, ultra-high-power electromagnetic systems, space exploration and climate-stabilization geophysics. Wood obtained his Ph.D. in astrophysics and planetary and space physics at UCLA in 1965, following receipt of bachelor's degrees in chemistry and math in 1962. He has held faculty and professional research staff appointments at the University of California (from which he retired after more than four decades in 2006) and is a Research Fellow at the Hoover Institution at Stanford University. He has advised the U.S. Government in many capacities, and has received a number of awards and honors from both government and professional bodies. Wood is the author, co-author or editor of more than 200 unclassified technical papers and books and more than 300 classified publications, and is named as an inventor on more than 200 patents and patents-pending.

Dr. Joan B. Woodard is Executive Vice President and Deputy Laboratories Director for Nuclear Weapons at Sandia National Laboratories. Sandia's role is to provide engineering support and design to the Nation's nuclear weapons stockpile, provide our customers with research, development, and testing services, and manufacture specialized non-nuclear products and components for national defense and security applications. The laboratories enable safe and secure deterrence through science, engineering, and management excellence. Prior to her current assignment, Dr. Woodard served as Executive Vice President and Deputy Director, responsible for Sandia's programs, operations, staff and facilities; developing policy and assuring implementation; and strategic planning. Her Sandia history began in 1974, and she rose through the ranks to become the Director of the Environmental Programs Center and the Director of the Product Realization Weapon Components Center; Vice President of the Energy & Environment Division and Vice President of the Energy Information and Infrastructure Technologies Division. Joan has been elected to the Phi Kappa Phi Honor Society and has served on numerous external panels and boards, including the Air Force Scientific Advisory Board, the National Academy of Sciences' Study on Science and Technology for Countering Terrorism, the Secretary of Energy's Nuclear Energy Research Advisory Council, the Congressional Commission on Electromagnetic Pulse, and the Intelligence Science Board. Joan has received many honors, including the Upward Mobility Award from the Society of Women Engineers, and was named as "One of Twenty Women to Watch in the New Millennium" by the Albuquerque Journal. She also received the Spirit of Achievement Award from National Jewish Hospital.

Dr. Michael J. Frankel is Executive Director of the EMP Commission and one of the Nation's leading experts on the effects of nuclear weapons. Formerly he served as Associate Director for Advanced Energetics and Nuclear Weapons, Office of the Deputy Undersecretary of Defense (S&T); Chief Scientist, Nuclear Phenomenology Division, Defense Threat Reduction Agency; Congressional Fellow, U.S. Senate; Chief Scientist, Strategic Defense Initiative Organization Lethality Program; and as a Research Physicist at the Naval Surface Warfare Center, White Oak. In prior government service, Dr. Frankel directed significant elements of the core national Nuclear Weapons Phenomenology program along with major WMD, Directed Energy, and Space System technology programs at the Defense Nuclear Agency while coordinating activities between the Military Services, National Laboratories, and industrial S&T organizations to address strategic defense technology needs. He has been an active participant in international scientific exchanges in his role as Executive Secretary for the U.S. – United Kingdom Joint Working Group under terms of the 1958 Atomic Treaty, and as Chairman of both the Novel Energetics and Hard Target Defeat working groups under the TTCP agreement with the UK, Australia, Canada and New Zealand. He has also delivered invited lectures, chaired national and international technical symposia, and published numerous articles in the professional scientific literature. He holds a Ph.D. in Theoretical Physics from New York University.

End Notes

[1] Perrow, Charles, *Normal Accidents,* Princeton University Press, Princeton, N.J., 1999.

[2] Neumann, Peter, *Computer-Related Risks*, Addison Wesley, Reading, Mass., 1995.

[3] Kohlberg, Clark, and Morrison, "Theoretical Considerations regarding the Interdependence between Power and Telecommunications," preprint, EMP Commission Staff Paper.

[4] Brown and Beyeler, "Infrastructure Interdependency Analysis of EMP Effects and Potential Economic Losses," EMP Commission Interdependencies Modeling and Simulation Workshop, Washington, D.C., June 2003.

[5] Wegleitner, Mark, Verizon, Senior Vice President, Broadband Packet Evolution, Technology, 2005.

[6] Executive Order 12472, April 3, 1984.

[7] http://www.globalsecurity.org/security/library/congress/2004_h/040302-libutti.htm.

[8] For example, Network Level EMP Effects Evaluation of the Primary PSN Toll-Level Networks, Office of the Manager: NCS, January 1994.

[9] http://www.y2k.gov/docs/infrastructure.htm.

[10] Aduskevicz, P., J. Condello, Capt. K. Burton, Review of Power Blackout on Telecom, NRIC, August 27, 2003, quarterly meeting.

[11] Convergence Working Group's final report, Impact of Network Convergence on NS/EP Telecommunications: Findings and Recommendations, February 2002.

[12] AT&T (Albert Lewis) correspondence with FCC, May 13, 1998; MCI (Bradley Stillman) correspondence with FCC, December 8, 1999.

[13] 47 CFR Part 215 designated the Executive Agent, NCS, as the focal point within the Federal Government for all EMP technical data and studies concerning telecommunications.

[14] Aduskevicz, P., J. Condello, Capt. K. Burton, Review of Power Blackout on Telecom, NRIC, August 27, 2003, quarterly meeting.

[15] Homeland Security Physical Security Recommendations for Council Approval, Letter to Richard C. Notebaert, March 5, 2003.

[16] United States, The President's National Security Telecommunications Advisory Committee, *Financial Services Risk Assessment Report* (Washington, 1997), 4.

[17] Ibid.

[18] "Firsts and Records," NYSE Euronext, New York Stock Exchange Euronext, http://www.nyse.com/about/history/1022221392987.html.

[19] Network General Corporation, Securities Industry Automation Corporation — SIAC: Sniffer Distributed, San Jose, 2005, 1.

[20] Ibid.

[21] Boston Options Exchange, Telecom Connections, August 3, 2003, http://www.bostonoptions.com/conn/tel.php.

[22] United States, The President's National Security Telecommunications Advisory Committee, *Financial Services Risk Assessment Report* (Washington, 1997), 47.

[23] ATM & Debit News, September 10, 2003, ATM & Debit News Survey Data Offers Insight into Debit Card and Network Trends in Its 2004 EFT Data Book, press release, http://www.sourcemedia.com/pressreleases/20030910ATM.html.

[24] MacAndrews, James J., and Simon M. Potter, "Liquidity Effects of the Events of September 11, 2001," Federal Reserve Bank of New York Economic Policy Review, November 2002.

[25] The Federal Reserve Board, the Office of the Comptroller of the Currency, and the Securities and Exchange Commission, *Interagency Paper on Sound Practices to Strengthen the Resilience of the U.S. Financial System* (Washington: GPO, 2002), 5.

[26] Systemic risk includes the risk that failure of one participant in a transfer system or financial market to meet its required obligations will cause other participants to be unable to meet their obligations when due, causing significant liquidity or credit problems or threatening the stability of financial markets. The use of the term "systemic risk" in this report is based on the international definition of systemic risk in payments and settlement systems provided in Committee on Payment and Settlement Systems, Bank for International Settlements, "A Glossary of Terms in Payment and Settlement Systems," 2001.

[27] "Federal Reserve Board Sponsorship for Priority Telecommunications Services of Organizations That Are Important to National Security/Emergency," *Federal Register*, 67:236 (December 9, 2002), 72958.

[28] Ibid.

[29] Ibid.

[30] Ibid.

[31] United States, Federal Reserve Board, *Federal Reserve Bulletin Statistical Supplement* (Washington: GPO, 2004), 15.

[32] Investment Company Institute, *2005 Investment Company Factbook,* 2005, http://www.ici.org/factbook.

[33] Klee, Elizabeth C., and Fabio M. Natalluci, "Profits and Balance Sheet Developments at U.S. Commercial Banks in 2004," *Federal Reserve Bulletin,* Spring 2005:144.

[34] United States Credit Union Statistics, Credit Union National Association, 2004, http://advice.cuna.org/download/us_totals.pdf.
[35] "Industry Overview: Investment Banking," Hovers, Inc., http://www.hoovers.com/investment-banking-/--ID 209--/free-ind-fr-profile-basic.xhtml.
[36] Ibid.
[37] Investment Company Institute, *2005 Investment Company Factbook,* 2005, 59, http://www.ici.org/factbook/pdf/05_fb_table01.pdf.
[38] Federal Reserve Board, http://www.federalreserve.gov/paymentsystems/coreprinciples/default.htm#fn12.
[39] Ibid.
[40] United States, Federal Reserve System, Analysis of Noncash Payments Trends in the United States: 2000–2003 (Washington: 2004), 5.
[41] SWIFT, *2005 Annual Report: Alternative Connectivity for CHIPS Reinforces Resilience*, http://www.swift.com/index.cfm?item_id=59677.
[42] SWIFT, *2004 Annual Report: SWIFTnet Now the Benefits Really Begin,* http://www.swift.com/index.cfm?item_id=56868.
[43] NASDAQ, *NASDAQ Announces Market Year-end Statistics for 2004,* http://ir.nasdaq.com/releasedetail. cfm?ReleaseID=177077.
[44] DTCC, *2004 Annual Report: What is a Quadrillion? 3*, http://www.dtcc.com/downloads/annuals/2004/2004_report.pdf.
[45] United States, The President's National Security Telecommunications Advisory Committee, *Financial Services Risk Assessment Report* (Washington, 1997), 40.
[46] Ibid.
[47] Ibid.
[48] National Academies of Science, *Making the Nation Safer: The Role of Science and Technology in Countering Terrorism* (Washington: National Academies Press, 2002), 137.
[49] Ibid.
[50] SWIFT, 2005 Annual Report: Alternative Connectivity for CHIPS Reinforces Resilience, http://www.swift.com/index.cfm?item_id=59677.
[51] Ibid.
[52] One Chicago (April 30, 2002), ONECHICAGO, Options Clearing Corporation and Chicago Mercantile Exchange, Inc., Sign Clearinghouse Agreements, press release, http://www.onechicago.com/060000_press_news/press_news_2002/04302002.html.
[53] "Digital Transactions News," *Digital Transactions*, January 6, 2005, MasterCard Worldwide, Digital Transactions, http://www.digitaltransactions.net/newsstory.cfm?newsid=466.
[54] Malphrus, Steve, Staff Director for Management, Federal Reserve Board, personal communication.
[55] Jackson, William D., *Homeland Security: Banking and Financial Infrastructure Continuity,* U.S. Congress, March 16, 2004, Congressional Research Service (Washington, 2004), 6, http://www.law.umaryland.edu/marshall/crsreports/crsdocuments/RL3187303162004.pdf.
[56] U.S. Security Exchange Commission, *Interagency Paper on Sound Practices to Strengthen the Resilience of the U.S. Financial System,* April, 2003.
[57] United States, The President's National Security Telecommunications Advisory Committee, *Financial Services Risk Assessment Report* (Washington, 1997), 27.
[58] National Academies of Science, *Making the Nation Safer: The Role of Science and Technology in Countering Terrorism* (Washington: National Academies Press, 2002), 137.
[59] Ibid.
[60] U.S. Security Exchange Commission, *Interagency Paper on Sound Practices to Strengthen the Resilience of the U.S. Financial System,* April 2003.
[61] Annual Energy Review 2006, International Energy Agency.
[62] Pipeline 101, http://www.pipeline101.com.
[63] National Petroleum Council, Securing Oil and Natural Gas Infrastructures in the New Economy, a Federal Advisory Committee to the Secretary of Energy, June 2001.
[64] Energy Information Administration, About Natural Gas, http://www.eia.doe.gov/pub/oil_gas/natural_gas/analysis_publications/ngpipeline/transsys_design.html.
[65] Energy Information Administration, Oil and Gas Code Field Master List, 2006.
[66] Natural Gas Processing Plants, 1995-2004 EIA 6/2006.
[67] National Petroleum Council, Securing Oil and Natural Gas Infrastructures in the New Economy. Not all interdependencies are shown.
[68] NSTAC Information Infrastructure Group Report, June 1999, http://www.ncs.gov/nstac/reports/1999/NSTAC22-IIG.pdf.
[69] Pipelines are sometimes associated with the transportation infrastructure but can be considered more usefully as part of the petroleum and natural gas infrastructures.

[70] The division of railroads into classes based on total operating revenue was a taxonomy defined by the Interstate Commerce Commission in the 1930s. The original threshold for a Class I railroad was $1 million. In 2006, Class I railroads were those with operating revenues exceeding $319.3 million. In North America, there are currently seven U.S. railroads that are defined as Class I, with an additional two Canadian railroads that would be considered Class I if U.S. definitions were applied. The old Class II and Class III designators are rarely used today. Instead, the Association of American Railroads speaks of regional railroads operating greater than 350 route-miles or generating more than $40 million revenue, local line haul carriers with less than 350 route-miles and generating less than $40 million revenue, and switching and terminal services carriers with highly localized functions, http://www.railswest.com/railtoday.html.

[71] "Tons originated" is a common term of art and index in the railroad industry used to track freight traffic volume. It is equal to the tons of traffic shipped by rail. Tons originated rail statistics are available from 1899.

[72] Association of American Railroads, http://www.aar.org.

[73] Some coal plants also can use natural gas, but this alternative fuel may not be available after an EMP attack. See Section 5, Petroleum and Natural Gas Infrastructures.

[74] Association of American Railroads, http://www.aar.org.

[75] Hansen, R.A., H. Schaer, D. Koenigstein, H. Hoitink, "A Methodology to Assess Exo-NEMP Impact on a Real System—Case Studies," EMC Symposium, Zurich, March 7 to 9, 1989. Reference describes EMP test of electro- mechanical locomotive belonging to Swiss Federal Railways.

[76] The Swiss executed both free-field (up to 25 kV/m) and current-injection (up to 2 kA) tests on a 4.6 MW, 80-ton electro-mechanical locomotive in both power-on and power-off configurations. During the free-field illumination, the test report states that "important analog/digital control electronics, deep inside the PC-boards, was repeatedly burnt out."

[77] Conversation with Colorado Springs lead traffic engineer.

[78] American Association of Port Authorities, http://www.aapa-ports.org.

[79] Ibid.

[80] Many satellites are likely to be unaffected by either EMP or by enhanced space radiation environment produced by a high-altitude burst (see Section 10 of this volume), but there may be some degradation as a result of vulnerabilities of receivers or ground stations.

[81] Bureau of Transportation Statistics.

[82] RTCA, Inc., is a not-for-profit corporation that develops recommendations regarding communications, navigation, surveillance, and air traffic management system issues.

[83] The FAA LINCS is a highly diverse private network constructed to meet specific requirements of a customer with critical mission requirements. The FAA LINCS is the most available private line network in the world with an off- backbone availability requirement of 99.8 percent. More than 21,000 circuits serve the entire network. More than 200 circuits form the LINCS backbone and satisfy diversity requirements of 99.999 percent availability. Despite natural disasters, major failures of public infrastructures, and the 2001 terrorist attacks, the FAA LINCS survived as designed, keeping the line of communication open between air traffic controllers and airplanes. In July 2002, the FAA initiated a substantial modernization of its telecommunications networks to meet its growing operational and mission support requirements and to provide enhanced security features. The new FTI Program is an integrated suite of products, services, and business practices that provide a common infrastructure supporting the National Airspace System (NAS) requirements for voice, data, and video services; improve visibility into network operations, service delivery status, and cost of services; and integrate new technologies as soon as they emerge. Reference: NSTAC Financial Services Task Force Report on Network Resilience, http://www.ncs.gov/nstac/nstac_ publications. html.

[84] Telecommunications Service Priority (TSP), http://tsp.ncs.gov.

[85] FEMA, Response and Recovery, Emergency Support Function No. 11 Food Annex, http://www. au. af. mil/au/awc/awcgate/frp/frpesf11.htm.

[86] Appendix B, Robert T. Stafford Disaster Relief and Emergency Assistance Act, P.L. 93-288, as amended (as of September 1, 1999), p. B-43, http://www.fema.gov/pdf/government/grant/pa/pagappb.pdf.

[87] National Emergency Number Association.

[88] Bureau of Labor Statistics. Frontline workers, including volunteers; excludes supervisory personnel.

[89] Rudman, Warren B., et al., *Emergency Responders: Drastically Underfunded, Dangerously Unprepared,* Council on Foreign Relations, 2003.

[90] E9-1-1 provides emergency services personnel with geographic location information on mobile callers.

[91] PSAP Enrollment in the TSP Program, http://www.nasna911.org/pdf/tsp-enroll-guide.pdf.

[92] National Communication System, http://www.ncs.gov/services.html.

[93] See also Section 6, Transportation Infrastructure.

[94] Burnett, John. "What Happened on New Orlean's Danziger Bridge?" http://www.npr.org/templates/story/story.php ?storyId=6063982.

[95] Radasky, William A., *The Threat of Intentional Electromagnetic Interference (IEMI) to Wired and Wireless Systems.* Metatech Corporation, Goleta, California, 162.

[96] Barnes, Paul R., *The Effects of Electromagnetic Pulse (EMP) on State and Local Radio Communications*, Oak Ridge National Laboratory, October 1973.

[97] Metropolitan Interoperability Radio System — Alexandria Site Description Document, *Advanced Generation of Interoperability for Law Enforcement (AGILE)*, Report No. TE-02-03, April 4, 2003.

[98] See Section 3, Telecommunications.

[99] Estimates based on a survey of local PSAPs, extrapolated to the entire country.

[100] Statistics obtained from the National Fire Protection Association.

[101] Report of the Commission to Assess United States National Security Space Management and Organization, January 11, 2001.

[102] The high-altitude test originally known as STARFISH was not successful. A second high-altitude test called STARFISH PRIME was successfully executed at a later date to obtain the sought-after data. In much of the literature describing the damage to satellites from this test, the name of the event is called STARFISH without the PRIME modifier. For the sake of brevity we also have dropped the modifier.

[103] Brown, W.L., W.N. Hess, and J.A. Van Allen, "Collected Papers on the Artificial Radiation Belt From the July 9, 1962, Nuclear Detonation," *Journal of Geophysical Research 68*, 605, 1963.

[104] Christofilos, N.C., Proceedings of the National Academy of Sciences, U.S. 45, 000, 1959.

[105] Van Allen, J.A., and L.A. Frank, "Radiation Around the Earth to a Radial Distance of 107,400 km," *Nature*, 183, 430, 1959.

[106] Schreiber, H., "Space Environments Analyst, Version 1.2," 1998 Space Electronics, Inc., Calculations using Space Radiation 4.0, Space Radiation Associates, Eugene, OR, 1998.

[107] Frederickson, A.R., "Radiation-Induced Dielectric Charging in Space Systems and Their Interactions with Earth's Space Environment," eds. H.B. Garrett and C.P. Pike, Progress in Astronautics and Aeronautics, vol. 71, AIAA, 1980.

[108] It is conventional (and useful) to describe the magnetic field lines on which electrons are trapped as belonging to numbered L-shells. The L-value of a field line is the distance (in Earth radii measured from the location of Earth's dipole field source) at which the field line intersects the magnetic equator. The inner belt peaks around L = 1.3, and the outer belt, near L = 4. Trapped electrons rapidly gyrate about the field lines, bounce along the field lines between mirror points, and drift around the Earth.

[109] As illustrated in figure 10-5, magnetic field lines that intersect the Earth at high northern and southern latitudes extend outward into space to relatively large distances. Conversely, magnetic field lines that intersect the Earth at low latitudes extend relatively short distances into space. Consequently, geomagnetically trapped electrons created by detonations at high latitudes can propagate along field lines out to very high altitudes where satellites orbit, whereas trapped electrons created by low-latitude bursts would be less likely to do so.

[110] While the use of twice the expected long-term-average exposure as a gauge of lifetime, as discussed here, is common practice, it relies entirely on total dose as a measure of radiation tolerance and ignores dose rate effects. Risks from circumstances involving nuclear detonations, where dose rates could be much larger than encountered under natural conditions, may be underestimated.

[111] The reader is reminded that our analysis deals only with collateral damage resulting from an EMP attack. Direct attack on satellites at any altitude, though serious, is not within the bounds of this analysis.

[112] Weenas, E.P., "Spacecraft Charging Effects on Satellites Following STARFISH Event," RE-78-2044-057, February 17, 1978.

[113] National Security and Homeland Security Presidential Directive, http://www.whitehouse.gov/news/releases/2007/05/20070509-12.html.

[114] Patrick, Eugene L., and William L. Vault, *Bioelectromagnetic Effects of the Electromatnetic Pulse (EMP)*, Adelphi, MD: Harry Diamond Laboratories, March 1990, pp. 6–7.

[115] Ibid, pp. 8–10.

[116] EMP Commission Staff Paper, Quick Look Pacemaker Assessment, December 2003.

[117] Sandia National Laboratory, EMP Commission-sponsored test.

[118] "The Great Northeast Blackout of 1965," http://www.ceet.niu.edu/faculty/vanmeer/outage.htm.

[119] Sigwart, Charles P., "Night of Terror," *Time*, July 25, 1977.

[120] "1977 New York Blackout," Blackout History Project, http://blackout.gmu.edu/events/tl1977.html.

[121] Kappenman, John G., "Geomagnetic Storms Can Threaten Electric Power Grid," *Earth in Space,* Vol. 9, No. 7, March 1997, pp.9-11. © 1997 American Geophysical Union. http://www.agu.org/sci_soc/eiskappenman.html.

[122] "Power failure brings New Zealand's largest city to standstill," CNN, http://www.cnn.com/WORLD/9802/24/nzealand.blackout/index.html.

[123] Adler, Jerry, et al, "The Day the Lights Went Out," *Newsweek*; August 25, 2003, Vol. 142, Issue 8, p. 44.

[124] Norris, et al, "60,000 Disaster Victims Speak: Part 1. An Empirical Review of the Empirical Literature, 1981-2001," *Psychiatry*, Fall 2002, 65, 3, Health Module.

[125] Barnes, Harper, "The Flood of 1993," *St. Louis Post-Dispatch*, July 25, 1993.

[126] Kendra, James, and Tricia Wachtendorf, "Elements of Resilience in the World Trade Center Attack," Disaster Research Center, 2001.

[127] Norris, et al, "60,000 Disaster Victims Speak: Part II. Summary and Implications of the Disaster Mental Health Research," *Psychiatry*, Fall 2002, 65, 3, Health Module.

[128] Warheit, G.J., "A note on natural disasters and civil disturbances: Similarities and differences." *Mass Emergencies*, 1, 1976, pp. 131-137.

[129] Ibid.

[130] The Federal Emergency Management Agency was an independent agency when the Commission was established; it is now a component within the Department of Homeland Security.

[131] This methodology is addressed in a Commission staff paper — Rob Mahoney, *Capabilities-Based Methodology for Assessing Potential Adversary Capabilities,* March 2004.

In: Vulnerabilities of the U.S. to an Electromagnetic Pulse Attack ISBN: 978-1-63484-477-2
Editor: Maryanne Schneider

Chapter 2

CRITICAL INFRASTRUCTURE PROTECTION: PRELIMINARY OBSERVATIONS ON DHS EFFORTS TO ADDRESS ELECTROMAGNETIC THREATS TO THE ELECTRIC GRID. STATEMENT OF CHRISTOPHER P. CURRIE, DIRECTOR, HOMELAND SECURITY AND JUSTICE, GOVERNMENT ACCOUNTABILITY OFFICE. HEARING ON "PROTECTING THE ELECTRIC GRID FROM THE POTENTIAL THREATS OF SOLAR STORMS AND ELECTROMAGNETIC PULSE"*

United States Government Accountability Office

WHY GAO DID THIS STUDY

The threat posed by an electromagnetic pulse (EMP) or solar weather event could have a debilitating impact on the nation's critical electrical infrastructure, as well as other key assets that depend on electricity. These events could lead to power outages over broad geographic areas for extended durations. Addressing these risks requires collaboration among multiple government and industry stakeholders; with DHS in the lead role for overall infrastructure protection efforts, working in coordination with DOE.

The EMP Commission, established by statute and comprised of subject matter experts, issued recommendations in 2008 addressing the preparation, protection and recovery of critical infrastructures against a possible EMP attack. The majority of these recommendations were made to DHS and DOE.

* This is an edited, reformatted and augmented version of a report, GAO-15-692T, issued by the U.S. Government Accountability Office. The report was originally presented as testimony July 22, 2015 before the Senate Committee on Homeland Security and Governmental Affairs.

This testimony is based on preliminary observations from GAO's ongoing review of DHS's efforts to address electromagnetic threats. Specifically, this testimony addresses the extent to which DHS has: (1) taken action to address recommendations from the 2008 EMP Commission Report and (2) coordinated with other principal federal agencies, such as DOE and industry stakeholders to mitigate risks to the electric grid from electromagnetic threats.

GAO reviewed EMP Commission recommendations and DHS program documents, and interviewed relevant stakeholders who provided insights on key issues and coordination activities with the federal government to address these threats.

WHAT GAO FOUND

As of July 2015, the Department of Homeland Security (DHS) reported taking several actions that could help address electromagnetic threats to the electric grid. GAO's preliminary analysis of DHS's actions indicates that they generally fell under four categories: (1) developing reports, (2) identifying mitigation efforts, (3) strategy development and planning, and (4) conducting exercises. For example:

- *Impacts of Severe Space Weather on the Electric Grid.* This 2011 report evaluated how previous solar storms have affected electric grids, and identified potential cost-effective mitigation equipment available to protect these grids, among other topics.
- *RecX.* In 2012, DHS Science &Technology partnered with industry to develop a prototype transformer that could significantly reduce the time to transport, install, and energize a transformer to aid recovery from power outages associated with transformer failures from several months to less than one week.

DHS reported its actions were not taken in response to the 2008 recommendations of the *Commission to Assess the Threat to the United States from Electromagnetic Pulse Attack* (EMP Commission). GAO also recognizes that DHS does not have a statutory obligation to specifically address the recommendations, but implementation of them could help mitigate electromagnetic impacts to the electric grid, such as helping to assure the protection of high-value transmission assets. Moreover, GAO's preliminary work suggests that DHS, in conjunction with the Department of Energy (DOE), has not fully addressed a key critical infrastructure protection responsibility—identification of clear internal agency roles and responsibilities related to addressing electromagnetic threats. For example, although DHS recognized one component as the lead for assessing solar weather risks, the component has not yet identified any specific roles related to collecting or analyzing risk information.

DHS has also coordinated with federal and industry stakeholders to address some, but not all risks to the electrical grid since the EMP Commission issued its recommendations. GAO preliminarily identified eight projects in which DHS coordinated with stakeholders to help protect the grid including developing plans to address long term power outages, participation in exercises, and research and development activities. Although these are positive steps, GAO's preliminary work indicates that DHS has not effectively coordinated with stakeholders to identify critical assets or collect necessary risk information, among other responsibilities. GAO will continue to assess the issues in this statement as it completes its work and will issue a report with the final results later this year.

* * *

Chairman Johnson, Ranking Member Carper, and Members of the Committee:

Thank you for the opportunity to discuss our work regarding the Department of Homeland Security's (DHS) efforts to address electromagnetic threats to the electric grid. The threat posed by an electromagnetic pulse (EMP) or solar weather event could have a debilitating impact on critical electrical infrastructure and communications systems, as well as other key assets and infrastructure that depend on electric utilities for power. EMP and solar weather events could potentially lead to power outages over broad geographic areas for extended durations, which experts have reported could result in severe economic disruption and significant impacts to public health and safety. Addressing these threats necessitates effective collaboration among multiple government agencies and industry partners and no singular federal program or entity has sole responsibility for addressing electromagnetic threats. However, the National Infrastructure Protection Plan (NIPP) outlines the roles and responsibilities of DHS and applicable sector-specific agencies for each of the 16 critical infrastructure sectors.[1] Within the NIPP framework, DHS has the lead role in coordinating the overall federal effort to promote the security and resilience of the nation's critical infrastructure. For the energy sector, which includes critical electrical infrastructure, the Department of Energy (DOE) is the sector-specific agency and shares responsibility with DHS.

In April 2008, the Commission to Assess the Threat to the United States from Electromagnetic Pulse Attack (EMP Commission) issued a report which included recommendations addressing the preparation, protection and recovery of U.S. critical infrastructures against a possible EMP attack.[2] The majority of these recommendations were made to DHS and to DOE. (See Appendix I for a summary of the 2008 EMP Commission recommendations addressing electrical infrastructure.)

My statement today is based on preliminary observations and analyses from our ongoing review of DHS's efforts to address electromagnetic threats.[3] Specifically, I will be discussing the extent to which DHS has: (1) taken actions to address recommendations from the 2008 EMP Commission Report, and (2) coordinated with other principal federal agencies and industry stakeholders to mitigate risks to the electric grid from electromagnetic threats.

To perform our ongoing work, we reviewed EMP Commission information and recommendations, as well as applicable laws and directives related to DHS's critical infrastructure protection responsibilities. To assess DHS actions to address electromagnetic threats, we reviewed DHS program documents, research reports, applicable risk assessments, and other supporting documentation such as program briefings and after action reports. Additional information on DHS coordination efforts was collected through interviews with multiple DHS components and other principal federal agencies addressing electromagnetic threats, as well as industry associations, subject matter experts from research organizations, product manufacturers, and electric utility operators. The non-federal entities we interviewed were identified, by federal officials and through our background research, as key stakeholders and subject matter experts within the electrical sector. While we intend to conduct additional interviews with industry stakeholders and researchers as part of our ongoing review, we believe the meetings conducted to date provided valuable insights regarding key issues and applicable coordination activities with the federal government to address electromagnetic threats.

We are conducting the work on which this statement is based in accordance with generally accepted government auditing standards. Those standards require that we plan and perform the audit to obtain sufficient, appropriate evidence to provide a reasonable basis for our findings and conclusions based on our audit objectives. We believe the evidence obtained provides a reasonable basis for our preliminary findings and conclusions based on our audit objectives.

BACKGROUND

An EMP is a burst of high power electromagnetic radiation resulting from the detonation of nuclear and non-nuclear devices that are designed to intentionally disrupt or destroy electronic equipment. EMP events may be further categorized into a number of different types, based on their specific source of initiation. The threat focused on primarily by the EMP Commission is the high-altitude EMP (HEMP). A HEMP event is caused by the detonation of a nuclear device at a high-altitude, about 40 to 400 kilometers, above the Earth's atmosphere.[4] A HEMP attack is not intended to cause direct physical impacts at the Earth's surface, such as injury or damage directly from heat or blast, but instead interacts with the atmosphere to create an intense electromagnetic energy field that can overload computer circuitry and could cause significant damage to critical electrical infrastructure.

In addition to manmade EMPs, naturally occurring solar weather events can also cause related electromagnetic impacts that can adversely affect components of the commercial electric grid. This type of event is commonly referred to as a geomagnetic disturbance (GMD).[5] In 1989, a GMD caused wide-scale impacts on the Hydro-Quebec power system in Canada which caused the electric grid to collapse within 92 seconds and left six million customers without power for 9 hours. As noted in Presidential Policy Directive 21 (PPD-21), energy sector infrastructure is uniquely critical due to the enabling functions it provides to other critical infrastructure sectors.[6] Given this interdependency, an EMP or major GMD event that disrupts the electric grid could also result in potential cascading impacts on fuel distribution, transportation systems, food and water supplies, and communications and equipment for emergency services, as well as other communication systems which utilize the civilian infrastructure. PPD-21 also recognizes that DHS has numerous responsibilities to protect critical infrastructure, including such things as analyzing threats to, vulnerabilities of, and potential consequences from all hazards on critical infrastructure.

Within DHS, the National Protection and Programs Directorate (NPPD) is responsible for working with public and industry infrastructure partners and leads the coordinated national effort to mitigate risk to the nation's infrastructure through the development and implementation of the infrastructure protection program. NPPD has two principal offices with responsibilities to facilitate protection of critical infrastructure that could be at risk from EMP and GMD events—the Office of Infrastructure Protection (IP) and the Office of Cyber Security and Communications (CS&C). In addition, DHS's Federal Emergency Management

Agency (FEMA) and Science and Technology Directorate (S&T) have roles related to addressing potential impacts to the electric grid, which could include EMP and GMD threats.

DOE also has a significant role as the sector-specific agency for the energy sector, which includes critical infrastructure and key resources related to electricity. For example, DOE is responsible for developing an Energy Sector Specific Plan—in collaboration with other stakeholders, including DHS—that applies the NIPP risk management model to critical infrastructure and key resources within the sector. Within DOE, the Office of Electricity Delivery and Energy Reliability leads national efforts to increase the security and reliability of the energy infrastructure and facilitate recovery from disruptions to the energy supply. DOE national laboratories also provide research support and technical expertise to federal and industry stakeholders regarding EMP and GMD impacts.

Other principal federal agencies working to address the threat of EMP and GMD include the Department of Defense (DOD) and the Federal Energy Regulatory Commission (FERC), as well as the National Oceanic and Atmospheric Administration (NOAA), and National Aeronautics and Space Administration (NASA).[7]

Electrical infrastructure is primarily operated by private industry which owns approximately 85 percent of the nation's critical electrical infrastructure. Industry entities are represented, in part, through membership in industry associations such as the American Public Power Association and the Edison Electric Institute. The North American Electric Reliability Corporation (NERC) also serves as the delegated authority to regulate the protection and improvement of the reliability and security of the electrical infrastructure.[8]

PRELIMINARY FINDINGS INDICATE THAT DHS ACTIONS TO ADDRESS ELECTROMAGNETIC THREATS WERE CONDUCTED INDEPENDENTLY OF THE EMP COMMISSION RECOMMENDATIONS

As of July 2015, DHS reported taking several actions that could help address electromagnetic threats to the electric grid, but these efforts were conducted independently of the 2008 EMP Commission recommendations. Our preliminary analysis of DHS's actions indicates that they generally fell under four categories of effort: (1) developing reports, (2) identifying mitigation efforts, (3) strategy development and planning, and (4) conducting training exercises.

Developing Reports

Since 2008, DHS has produced three reports that specifically address electromagnetic threats to the electric grid. Below is a summary of each report.

- *Electromagnetic Pulse Impacts on Extra High Voltage Power Transformers.*[9] This 2010 report analyzed the potential impact of an EMP on extra high voltage transformers—focusing primarily on transformer equipment designs and identifying specific mitigation efforts such as blocking devices that minimize the impact of geomagnetically induced currents (GIC) on the electric grid.[10] The report concluded that the similarity of EMP effects, regardless of source, indicates that geomagnetic storms provide a useful basis for transformer impact analysis and that selective installation of blocking devices would minimize the impacts of GIC on transformers, among other findings.

- *Impacts of Severe Space Weather on the Electric Grid.*[11] This 2011 report assessed the impacts of space weather on the electric grid, seeking to understand how previous solar storms have affected some power grids, and what cost-effective mitigation efforts are available to protect the electric grid, among other topics. Some of the key findings and recommendations include the need for a rigorous risk assessment to determine how plausible a worse-case scenario may be and additional research to better understand how transformers may be impacted by electromagnetic threats. This report also recommended installation of blocking devices to minimize the impacts of GIC.
- *Sector Resilience Report: Electric Power Delivery.*[12] This 2014 report summarizes an analysis of key electric power dependencies and interdependencies, such as communications, transportation, and other lifeline infrastructure systems. The report included an assessment of, and best practices for, improving infrastructure resilience such as: modeling to identify potential vulnerabilities, conducting a cost-benefit analysis of alternative, technology-based options, and installing protective measures and hardening at-risk equipment, among others.

Identifying Mitigation Efforts

DHS identified two specific efforts implemented since 2008 that could help to mitigate electromagnetic impacts to the electric grid. They are: (1) Recovery Transformer Project (RecX), and (2) Cyber Emergency Response Team.

- *RecX.* In 2012, S&T partnered with industry to develop a prototype transformer that could significantly reduce the time to transport, install, and energize a transformer to aid recovery from power outages associated with transformer failures from several months to less than one week. S&T, along with industry partners, demonstrated the RecX prototype for 2.5 years, ending in September 2014. DHS reported that RecX proved to be successful in an operational environment and has the capacity to reduce the impact of power outages.
- *Cyber Emergency Response Team.* CS&C operates the Industrial Control Systems-Cyber Emergency Response Team to assist critical infrastructure owners in the 16 sectors, including the energy sector, to improve overall cybersecurity posture of their control systems.[13] Industrial control systems are among the types of critical electrical infrastructure that could be impacted in the event of an EMP attack.

Strategy Development and Planning

DHS has taken actions to support the development of two key strategies and plans that could help to address electromagnetic threats. These include areas: 1) Power Outage Incident Annex, and 2) the National Space Weather Strategy.

- *Power Outage Incident Annex.* In 2014, FEMA began developing a Power Outage Incident Annex (incident annex) to provide incident-specific information, which supplements the National Response Framework.[14] According to FEMA officials, the incident annex will describe the process and organizational constructs that the federal government will utilize to respond to and recover from loss of power resulting from deliberate acts of terrorism or natural disasters. Among other tasks, the incident annex is designed to identify key federal government capabilities and resources, prioritize core capabilities, and outline response and recovery resource requirements. FEMA officials reported that the incident annex is scheduled to be completed by October 2015.
- *National Space Weather Strategy.* In collaboration with the White House Office of Science and Technology Policy and NOAA, DHS has been working since 2014 to help develop a National Space Weather Strategy. As a co-chair of the Space Weather Operations, Research and Mitigation Task Force, DHS is in the process of developing a strategy to achieve several goals, including efforts to establish benchmarks for space weather events, improve protection and mitigation efforts, and improve assessment, modeling, and prediction of impacts on critical infrastructure, among other goals. According to officials at S&T, a draft of the National Space Weather Strategy is currently being updated to incorporate stakeholder comments and is scheduled to be completed in September 2015.

DHS has also conducted two training exercises that could help address the potential impact of power outages caused by electromagnetic events, GridEx II and Eagle Horizon.[15]

- *GridEx II.* In November 2013, DHS, along with the Federal Bureau of Investigation, DOE, and other relevant government agencies, participated in an industry-wide exercise assessing the readiness of the electricity industry to respond to a physical or cyber attack on the bulk power system. The key goals of GridEx II were to review existing command, control, and communication plans and tools, incorporate lessons learned from a previous exercise, and to identify potential improvements in cyber and physical security plans and programs. Upon completing the exercises, participants identified key lessons learned, which included the need for enhanced information sharing, and clarification of roles and responsibilities during a physical or cyber attack.
- *Eagle Horizon.* Since 2004, FEMA has conducted a mandatory, annual continuity exercise for all federal executive branch departments and agencies to ensure the preservation and continuing performance of essential functions. Key objectives of the training exercise include: assessing the implementation of continuity plans, demonstrating communication capabilities, and examining broader national continuity capabilities with state, local, and private sector partners.

For our ongoing review, DHS did not identify its actions as specifically responsive to the EMP Commission's recommendations; nonetheless, some of the actions DHS has taken since 2008 could help to mitigate some electromagnetic impacts to the electric grid. For example, the three identified reports provide some insights on how the electric grid may be impacted by electromagnetic threats. Additionally, the RecX project provided a functional prototype that may facilitate industry efforts to further develop more mobile transformers and assist with recovery efforts in the event of an electromagnetic attack on the electric grid. Similarly, DHS planning efforts to develop the power outage incident annex and space weather strategy are also steps that could help to mitigate the negative effects of an electromagnetic threat to the electric grid by improving critical planning and response efforts.

While DHS has taken several positive steps to address electromagnetic threats to the electric grid since the EMP Commission issued its recommendations in 2008, our preliminary analysis indicates that these actions may fall short of the expectations for DHS regarding overall responsibilities to oversee and coordinate national efforts to protect critical electrical infrastructure, consistent with PPD-21 and the NIPP. For example, DHS's efforts to clearly identify agency roles and responsibilities to date have been limited. Specifically, DHS has had difficulty identifying the relevant DHS components, officials, or ongoing internal DHS activities with an EMP nexus. For example, DHS officials were unable to determine internally which component would serve as the lead—S&T or NPPD—in regards to addressing EMP threats. In addition, NPPD has not yet identified its specific roles and activities in addressing electromagnetic threats even though it has been identified by the DHS Office of Policy as the proposed risk analysis "owner" relative to space weather threats.[16]

We recognize that DHS does not have a statutory obligation to address the specific recommendations of the EMP Commission and many of these recommendations were also directed to DOE. Nevertheless, we believe that implementation of them could help mitigate electromagnetic impacts to the electric grid, such as helping to assure the protection of high-value transmission assets. Moreover, PPD-21 articulates DHS's roles and responsibilities to safeguard the nation's critical infrastructure, which are consistent with such recommendations. For example, PPD-21 states that DHS, in carrying out its responsibilities under the Homeland Security Act of 2002, as amended, is to, among other things, evaluate national capabilities, opportunities, and challenges in protecting critical infrastructure; analyze threats to, vulnerabilities of, and potential consequences from all hazards on critical infrastructure; identify security and resilience functions that are necessary for effective stakeholder engagement with all critical infrastructure sectors; integrate and coordinate federal cross-sector security and resilience activities; and identify and analyze key interdependencies among critical infrastructure sectors. Moreover, PPD-21 calls for DHS to specifically consider sector dependencies on energy and communications systems, and identify pre-event and mitigation measures or alternate capabilities during disruptions to those systems in updating the NIPP. To date, our preliminary analysis suggests that DHS has not fully addressed some key responsibilities related to effectively preparing for and responding to electromagnetic threats to the electric grid, in conjunction with DOE as the sector-specific agency for the energy sector, which is responsible for critical electrical infrastructure. Specifically, DHS did not identify any efforts it conducted to support the identification of key electrical infrastructure assets or assess cross-sector dependencies on these assets, for which DHS would be expected to play a key role. According to officials within NPPD and the DHS Office of Policy, factors such as competing priorities and a focus on all hazards may contribute to limited efforts being taken by DHS to specifically address electromagnetic threats. We will continue to assess the extent to which DHS's efforts align with the EMP Commission recommendations as well as the extent to which DHS's current and planned actions align with its own risk management framework, as identified in the NIPP, as we complete our work. We will report our final results later this year.

PRELIMINARY ANALYSIS INDICATES DHS HAS NOT FULLY COORDINATED WITH STAKEHOLDERS TO ADDRESS SOME RISKS TO THE ELECTRIC GRID

Our preliminary analysis indicates that since the EMP Commission issued its recommendations in 2008, DHS has coordinated with federal and industry stakeholders to address some, but not all risks to the electric grid. Specifically, DHS has not fully coordinated with stakeholders in certain areas such as identifying critical assets or collecting information necessary to assess electromagnetic risks. Our preliminary work has identified eight projects in which DHS coordinated with other federal agencies or industry to help protect the electric grid. These projects encompass a range of different protective efforts, including the development of plans to address long term power outages, participation in exercises, and research and development activities which address the resiliency of electrical infrastructure (See Appendix II for a list of projects we identified.)

Four of the eight projects we identified were initiated within the past 2 years and three specifically address the risks associated with an EMP or GMD event.[17] The three EMP or GMD-related projects include 1) participation in a White House Task Force to support development of an interagency space weather action plan; 2) collaboration with NASA to develop precise, localized forecasts that can help utilities better respond to solar weather events; and 3) development of EMP protection guidelines for critical equipment, facilities, and communications/data centers.

In addition to the specific projects identified above, DHS also coordinates with sector stakeholders through the Energy Sector Government Coordinating Council (EGCC)—which it co-chairs with DOE—and the Electricity Subsector Coordinating Council (ESCC) through the Critical Infrastructure Partnership Advisory Council. While federal officials generally identified that EMP and GMD issues have been discussed via these groups in recent years, they noted that the EMP threat has not been an area of particular focus.

Although DHS participation in the identified projects is a positive step to help mitigate some potential impacts of electromagnetic threats, our preliminary work suggests that DHS has not fully coordinated with stakeholders in other areas to help facilitate EMP and GMD protective efforts. Specifically, our preliminary analysis indicates that DHS has not fully coordinated with stakeholders to address electromagnetic threats to the electric grid in the following areas:

Providing threat information. DHS has not identified any efforts to specifically provide EMP-related threat information to industry stakeholders. Industry officials we spoke with generally stated that they do not have sufficient threat information to determine the extent to which specific actions should be taken to mitigate the effects of an EMP event. Whereas industry officials reported having a greater understanding of the potential likelihood of a major GMD caused by solar weather, they noted that applicable EMP threat briefings by DOD or DHS could help them to better justify to their management or stockholders the level of investment required to take protective actions. According to the Quadrennial Energy Review, incomplete or ambiguous threat information may lead to inconsistency in physical security among grid owners, inefficient spending on security measures, or deployment of security measures against the wrong threat.[18]

This concern generally aligns with previous work related to cyber threats in which we reported that federal partners' efforts to share information did not consistently meet industry's expectations, in part, due to restrictions on the threat information that can be shared with industry partners.[19] DHS generally concurred with our prior recommendations directed at strengthening its partnership and information-sharing efforts, and has since taken steps to enhance its information sharing activities, including granting security clearances, and establishing a secure mechanism to share cyber threat information. We will continue to assess DHS's actions regarding providing threat information on EMP as part of our ongoing work.

Identifying key infrastructure assets. Our preliminary analysis indicates that DHS and DOE have not taken action to identify the most critical substations and transformers on the electric grid. According to the NIPP risk management framework, such information is important to better understand system dependencies and cascading impacts, as well as help determine priorities for collecting additional information on specific asset vulnerabilities or potential mitigation actions.

According to the 2008 EMP Commission report, government entities, such as DHS and DOE, must identify the specific electrical assets that are critical to remain in service or that can be restored within hours following an EMP attack. Protection of these assets may be necessary to ensure the continuation of emergency response and recovery functions.

As part of our ongoing work, we will continue to assess actions by DHS and other federal agencies regarding the identification of key infrastructure assets.

Collecting risk information. DHS has not fully leveraged existing programs or utilized collaboration opportunities with federal partners to collect additional vulnerability and consequence information related to potential impacts to the electric grid. For example, DHS-IP has not fully leveraged the Infrastructure Survey Tool and Regional Resiliency Assessment Program (RRAP) to help collect additional information related to infrastructure vulnerabilities and impacts related to electromagnetic threats.[20] As we have concluded previously, coordination with other federal partners may also help ensure an integrated approach to vulnerability assessment activities.[21] For example, DHS has also not fully leveraged other agency efforts such as DOD's Defense Critical Infrastructure Protection program which could provide useful information about potential consequences of electric grid failure. According to the NIPP, to assess risk effectively, critical infrastructure partners—including owners and operators, sector councils, and government agencies—need timely, reliable, and actionable information regarding threats, vulnerabilities, and consequences. As part of our ongoing work, we will continue to assess actions by DHS and other federal agencies regarding the collection of applicable risk information.

Engaging with industry to identify research priorities and funding mechanisms. Enhanced collaboration among federal and industry partners is critical to help identify and address key research gaps and priorities, and leverage available funding mechanisms. Our preliminary analysis identified two areas—assessing transformer impacts and development of mitigation tools—where DHS has not fully pursued opportunities to collaborate with federal and industry stakeholders on research, testing and identifying funding sources that could help facilitate efforts to address electromagnetic threats to the electric grid. With respect to transformer impacts, industry and government officials identified the need for additional modeling and assessment as the most critical research gap. For example, the 2012 NERC GMD Task Force found that modeling the effects of GIC flows on transformers during a GMD event is not sufficiently developed. Stakeholders also noted that additional action is needed for evaluating and testing equipment that could help mitigate electromagnetic impacts to key infrastructure assets. Specifically, stakeholders identified that there are limited sites available for large-scale testing, and opportunities may exist to further leverage DOE research laboratories and other federal resources, including potential funding mechanisms.

In our ongoing review, we will continue to evaluate federal and industry actions to determine where specific coordination efforts could be improved and we will report the final results later this year.

Chairman Johnson, Ranking Member Carper and Members of the Committee, this completes my prepared statement. I would be pleased to respond to any questions that you may have at this time.

APPENDIX I: SUMMARY OF 2008 EMP COMMISSION RECOMMENDATIONS ADDRESSING ELECTRICAL INFRASTRUCTURE

Table 1. Summary of 2008 Electromagnetic Pulse (EMP) Commission Recommendations Addressing Electrical Infrastructure

EMP Commission Recommendations[a]
1. The Commission recommends research be conducted to better understand infrastructure system interdependencies and interactions, along with the effects of various EMP attack scenarios. In particular, the Commission recommended that such research include a strong component of interdependency modeling. Funding could be directed through a number of avenues, including the Department of Homeland Security (DHS) and National Science Foundation.
2. Expand activities to address the vulnerability of Supervisory Control and Data Acquisition (SCADA) systems to other forms of electronic assault, such as EMP.
3.It is vital that DHS, as early as practicable, make clear its authority and responsibility to respond to an EMP attack and delineate the responsibilities and functioning interfaces with all other governmental institutions with individual jurisdictions over the broad and diverse electric power system. This is necessary for private industry and individuals to act to carry out the necessary protections assigned to them and to sort out liability and funding responsibility.
4. DHS particularly needs to interact with the Federal Energy Regulatory Commission (FERC), North American Electric Reliability Corporation (NERC), state regulatory bodies, other governmental institutions at all levels, and industry in defining liability and funding relative to private and government facilities, such as independent power plants, to contribute their capability in a time of national need, yet not interfere with market creation and operation to the maximum extent practical.
5. DHS must establish the methods and systems that allow it to know, on a continuous basis, the state of theinfrastructure, its topology, and key elements. Testing standards and measurable improvement metrics should be defined as early as possible and kept up to date.

Table 1. (Continued)

EMP Commission Recommendations[a]
6. Working closely with industry and private institutions, DHS should provide for the necessary capability to control the system in order to minimize self-destruction in the event of an EMP attack and to recover as rapidly and effectively as possible.
7.DHS and DOE must utilize industry and other governmental institutions to assure the most cost effectiveoutcome occurs and that it does so more rapidly than otherwise possible. In many instances, these initiatives are extensions or expansions of existing procedures and systems such as those of NERC. Separate recommended initiatives are listed below. a) Understand system and network level vulnerabilities, including cascading effects b) Evaluate and implement quick fixes c) Develop national and regional restoration plans d) Assure availability of replacement equipment e) Assure availability of critical communications channels f) Expand and extend emergency power supplies g) Extend black start capability h) Prioritize and protect critical nodes i) Expand and assure intelligent islanding capability j) Assure protection of high-value generation assets k) Assure protection of high-value transmission assets l) Assure sufficient numbers of adequately trained recovery personnel m) Simulate, train, exercise, and test the recovery plan n) Develop and deploy system test standards and equipment o) Establish installation standards

Source: 2008 EMP Commission recommendations. | GAO-15-692T.

Notes:

[a]The EMP Commission recommendations cited above capture two key areas: infrastructure commonalities and the electric power sector which both have a nexus to electrical infrastructure. The Commission also made recommendations addressing potential EMP impacts affecting other infrastructure sectors, such as telecommunications, banking, and emergency services, among others.

APPENDIX II: DHS COORDINATION ON ACTIVITIES TO ADDRESS ELECTROMAGNETIC IMPACTS TO THE ELECTRIC GRID

Table 1. Department of Homeland Security (DHS) Coordination on Activities to Address Electromagnetic Impacts to the Electric Grid

Activity	Status	DHS Role	Focus Area (Electromagnetic or All Hazards)
Grid Security Exercise (GridEx II)[a]	Completed–2013	Participant	All-Hazards
Recovery Transformer (RecX)[b]	Completed–2014	Lead	All-Hazards
Resilient Electric Grid (REG)[c]	Ongoing	Lead	All-Hazards
Power Outage Incident Annex[d]	Ongoing	Lead	All-Hazards

Activity	Status	DHS Role	Focus Area (Electromagnetic or All Hazards)
National Emergency Communications Plan (2014 Update)[e]	Ongoing	Lead	All-Hazards
Space Weather Operations, Research and Mitigation Task Force[f]	Ongoing	Participant	Electromagnetic
Solar Storm Mitigationg	Ongoing	Co-Lead	Electromagnetic
EMP Protection Guidelines (Version 6.0)[h]	Ongoing	Lead	Electromagnetic

Source: GAO analysis of DHS activities addressing threats to the electric grid. | GAO-15-692T.

Notes:

[a] The North American Electric Reliability Corporation (NERC) conducted the second industry-wide Grid Security Exercise in 2013. The exercise simulated a coordinated cyber and physical attack on the electric grid and tested the response capability of government and industry stakeholders.

[b] The Recovery Transformer program was a partnership between DHS Science &Technology (S&T) and the electric industry to develop a prototype transformer that could be transported, installed, and energized in a shorter timeframe to aid recovery from power outages associated with transformer failures from several months to less than one week.

[c] The Resilient Grid program is being coordinated through S&T. Specifically, S&T is developing a superconducting cable that would allow substations to interconnect and share power while eliminating the risk of cascading fault currents. The cable enhances the flexibility and resiliency of the electric grid.

[d] FEMA is coordinating the development of a Power Outage Incident Annex (POIA). The POIA will describe the process and organizational constructs that the federal government will utilize to respond to and recover from loss of power resulting from deliberate acts of terrorism or natural disasters, including electromagnetic pulse (EMP) and space weather events.

[e] DHS developed the National Emergency Communications Plan to coordinate emergency communication across all levels of government, the private sector, and the nongovernmental sector.

[f] The Space Weather Operations, Research and Mitigation Task Force operates under the direction of the White House National Science and Technology Council and is working to finalize two key products: the National Space Weather Strategy, and the Space Weather Action Plan.

[g] The Solar Storm Mitigation project is being coordinated through S&T. For additional information, see dhs.gov/science-and-technology/solar-storm-mitigation.

[h] Development of EMP Protection guidelines is led by the Federal Executive Branch Continuity Communications Managers Group (CCMG). The guidelines include four levels of protection and are based on using specific devices such as EMP-capable surge arresters on power cords to mitigate EMP vulnerabilities.

End Notes

[1] See DHS, National Infrastructure Protection Plan, Partnering for Critical Infrastructure Security and Resilience (Washington, D.C.: December 2013). Sector-specific agencies are the federal departments and agencies responsible for providing institutional knowledge and specialized expertise, as well as leading, facilitating, or supporting the security and resilience programs and associated activities of its designated critical infrastructure sector in the all- hazards environment.

[2] Established pursuant to the fiscal year 2001 National Defense Authorization Act, the EMP Commission was responsible for assessing: 1) the nature and magnitude of potential high-altitude EMP threats to the United States; 2) the vulnerability of U.S. military and civilian systems to an EMP attack in terms of emergency preparedness; 3) the capability of the U.S. to repair and recover from damage inflicted by an EMP attack; and 4) the feasibility and cost of hardening select military and civilian systems against EMP attack. See Pub. L. No. 106-398, §§ 1401-09, 114 Stat. 1654, 1654A-345-348 (2000). See also Pub. L. No. 109-163, § 1052, 119 Stat. 3136, 3434-35 (2006) (reestablishing the EMP Commission to continue its efforts to monitor, investigate, make recommendations, and report to Congress on the evolving threat to the U.S. in the event of an EMP attack resulting from the detonation of a nuclear weapon or weapons at high altitude). See also Pub. L. No. 110-181, Div. A, § 1075 122 Stat. 3, 333 (2008) (providing, among other things, that the Commission and the

Secretary of Homeland Security shall jointly ensure that the work of the Commission with respect to EMP attack on electricity infrastructure, and protection against such attack, is coordinated with DHS efforts on such matters).

[3] We are conducting this work at the request of the Senate Committee on Homeland Security and Governmental Affairs and the House Committee on Homeland Security. Two individual Members of Congress are also requesters for this work.

[4] EMP events may also be initiated by other means. A Source Region EMP (in this context) is an electromagnetic pulse created when a nuclear weapon detonates at low altitude (surface or near-surface detonation). The electromagnetic field is small compared to that from HEMP and it affects a smaller geographic area. A System Generated EMP occurs when a nuclear weapon detonates above the atmosphere and sends out damaging x-rays that strike space systems, such as satellites. A Non-Nuclear EMP is a product of radio frequency weapons, which are devices that produce electromagnetic energy to burn out or disrupt components, systems, and networks. In addition, an intentional electromagnetic interference (IEMI) is the non-explosive, non-nuclear intentional generation of intense electromagnetic fields, which are directed to a target by an antenna. The electromagnetic fields are used to purposefully disrupt or confuse the targeted electronics.

[5] According to the National Oceanic and Atmospheric Administration's (NOAA) Space Weather Prediction Center, a GMD event is a major disturbance of the Earth's magnetosphere that occurs when there is an exchange of energy from the solar wind into the space environment surrounding Earth. In addition, larger GMD's are generally associated with solar coronal mass ejections (CME), which are explosions of magnetic field and plasma from the Sun's corona. A CME moves outward from the sun through solar wind to reach Earth within 18-96 hours, roughly 1-4 days after a CME.

[6] Presidential Policy Directive/PPD-21—Critical Infrastructure Security and Resilience (Feb. 12, 2013).

[7] FERC is an independent federal agency that regulates the interstate transmission of electricity, natural gas, and oil, and oversees the reliability of high-voltage interstate transmission systems, among other responsibilities. NOAA operates the Space Weather Prediction Center – a 24/7 space weather monitoring facility that provides alerts and warnings to applicable federal entities, emergency management personnel, and other affected parties, including operators of electric utilities.

[8] The North American Electric Reliability Corporation is a not-for-profit international regulatory organization whose mission is to ensure the reliability of the bulk power system in North America and is subject to oversight by FERC and governmental authorities in Canada.

[9] Department of Homeland Security, Electromagnetic Pulse (EMP) Impacts on Extra High Voltage Power Transformers, Rev. 2, April 2010.

[10] GIC are electric currents induced on a power system caused by fluctuations in the earth's magnetic field. Large-scale fluctuations in the earth's magnetic field that cause damaging GIC are usually caused by solar storms.

[11] MITRE, Impacts of Severe Space Weather on the Electric Grid, JSR-11-320 (McLean, VA: November 2011).

[12] Department of Homeland Security, Sector Resilience Report: Electric Power Delivery, June 11, 2014.

[13] Industrial control system is a general term that encompasses several types of control systems, including Supervisory Control and Data Acquisitions (SCADA) systems. SCADA systems are used extensively in critical infrastructure applications such as electrical transmission and distribution, water management, and oil and gas pipelines.

[14] The National Response Framework is a guide to how the nation responds to disasters and emergencies of all types and describes the principles, roles and responsibilities, and coordinating structures for delivering the core capabilities required to save lives, protect property and the environment, stabilize communities, and meet basic human needs following an incident.

[15] In commenting on information contained in a draft of this statement, an official with S&T noted that DHS participated in a third exercise addressing risks to the electric grid – Secure Grid 2011. According to this official, the exercise was conducted at National Defense University, was jointly funded by DHS, DOD, and DOE, and included participation with multiple federal agencies and industry representatives. Our ongoing work will review information on this exercise which DHS had not previously identified as an EMP-related action.

[16] According to the DHS Office of Policy, space weather has been included as an identified risk event since the initial Strategic National Risk Assessment was conducted in 2011, and more recently was included in the Homeland Security National Risk Characterization, which serves to identify the specific natural hazards to be assessed as part of the 2014 Quadrennial Homeland Security Review (QHSR). Every four years the Secretary is to complete a QHSR—a comprehensive examination of the homeland security strategy of the nation that is to include recommendations regarding the long-term strategy and priorities of the nation for homeland security

and guidance on the programs, assets, capabilities, budget, policies and authorities of the department. See 6 U.S.C. § 347.

[17] Although each of the eight projects identified may facilitate some level of risk reduction to electrical infrastructure, we identified three projects that were specifically initiated to address the unique causes or vulnerabilities associated with electromagnetic events.

[18] See Presidential Memorandum—Establishing a Quadrennial Energy Review (Jan. 9, 2014). The initial Quadrennial Energy Review was released on April 21, 2015.

[19] GAO, Critical Infrastructure Protection, Observations on Key Factors in DHS's Implementation of its Partnership Approach, GAO-14-464T (Washington, D.C.: March 6, 2014).

[20] The Infrastructure Survey Tool is a voluntary, web-based vulnerability survey conducted by DHS protective security advisors to identify and document the overall security and resilience of a facility. The RRAP is an analysis of infrastructure clusters and systems in specific geographic areas or regions. Using the RRAP, DHS examines vulnerabilities, threats, and potential consequences to identify (1) dependencies and interdependencies among the assets that participate in the RRAP, (2) cascading effects resulting from an all-hazards disruption of these assets or the region, (3) characteristics that make the assets and the region resilient, and (4) any resilience gaps that may hinder rapid recovery from disruptions.

[21] GAO, Critical Infrastructure Protection: DHS Action Needed to Enhance Integration and Coordination of Vulnerability Assessment Efforts, GAO-14-507 (Washington, D.C., Sept. 14, 2014). DHS concurred with our recommendation that DHS identify and analyze key critical infrastructure (CI) security-related assessment tools and methods used or offered by SSAs and provide guidance for what areas should be included in vulnerability assessments of CI that can be used by DHS and other CI partners in an integrated and coordinated manner.

INDEX

C

D

E

F

G

H

I

J

K

L

M

N

Q

R

T